Henry Van Brunt

Greek Lines and Other Architectural Essays

Henry Van Brunt

Greek Lines and Other Architectural Essays

ISBN/EAN: 9783337311896

Printed in Europe, USA, Canada, Australia, Japan

Cover: Foto ©berggeist007 / pixelio.de

More available books at **www.hansebooks.com**

GREEK LINES

AND OTHER ARCHITECTURAL ESSAYS

BY

HENRY VAN BRUNT
FELLOW OF THE AMERICAN INSTITUTE OF ARCHITECTS

BOSTON AND NEW YORK
HOUGHTON, MIFFLIN AND COMPANY
The Riverside Press, Cambridge
1893

To
WILLIAM ROBERT WARE,
THIS BOOK,
AN INADEQUATE RESULT OF MANY FRUITFUL YEARS OF
PERSONAL AND PROFESSIONAL RELATIONS WITH
HIM, IS AFFECTIONATELY INSCRIBED
BY THE AUTHOR.

PREFACE.

THIS little book contains a selection of architectural essays which have been written at intervals during an active professional career of more than thirty years. The kindly indulgence with which they have been received by my brethren in the craft, and the friendly curiosity with which they have been regarded by the laity, may be attributed perhaps mainly to the fact that, unlike most literary expositions of the subject, they have been developed rather from practice than from theory, and that they had the advantage of that sort of sincerity which is the natural product of conviction rather than sentiment. I have been encouraged to believe that it might be possible so to collate a few of these papers as to form a sequence, in which, from first to last, there should be evident a more or less orderly development of thought. To this end the earlier essays in especial have been largely rewritten, and all have been, so far as practicable, brought to date. The essay on Blois is the only one of the series which now makes its first appeal to the public. Perhaps, in their present consecutive form, these papers may have enough of symmetrical unity to secure for the architectural matters of which they treat a more intelligent popular interest than could be obtained by any sort of disconnected publication.

H. V. B.

Kansas City, Mo., July, 1893.

CONTENTS

	PAGE
GREEK LINES AND THEIR INFLUENCE ON MODERN ARCHITECTURE	1
THE GROWTH OF CONSCIENCE IN MODERN DECORATIVE ART	92
HISTORICAL ARCHITECTURE, AND THE INFLUENCE OF THE PERSONAL ELEMENT UPON IT	134
THE ROYAL CHÂTEAU OF BLOIS	164
THE PRESENT STATE OF ARCHITECTURE	202
ARCHITECTURE AND POETRY	234

LIST OF ILLUSTRATIONS

	PAGE
THE EGYPTIAN LOTUS	17
THE THREE TYPICAL LINES	24
GREEK DORIC CAPITALS	33
THE IONIC CAPITAL OF THE ERECTHEUM	38
THE GREEK AND ROMAN CYMA	49
THE ROMAN IONIC CAPITAL	52
THE CHÂTEAU OF BLOIS.	
General Plan	166
Pavilion of Louis XII., Exterior Façade	168
Pavilion of Louis XII., Detail of Exterior Façade	170
Pavilion of Louis XII., Interior Façade	172
Pavilion of Francis I., Interior Façade	184
Pavilion of Francis I., Grand Staircase	188
Pavilion of Francis I., Exterior Façade	192
DOORWAY OF ST. TROPHÊME AT ARLES	234

GREEK LINES

AND

OTHER ARCHITECTURAL ESSAYS.

GREEK LINES AND THEIR INFLUENCE ON MODERN ARCHITECTURE.

I.

THE material beauty of the world, as exhibited in the manifold forms, colors, sounds, perfumes, motions of nature, is created for a nobler purpose than only to delight the senses and please the æsthetic faculties. When, according to the suggestions of our merely human passions and instincts, we ease our hearts of love by heaping treasures and the choicest gifts of fancy in the laps of those whom we most dearly cherish, we take no credit to ourselves for such precious prodigalities; for they are the inevitable and disinterested outpourings of affection. They are received as such. When, therefore, we see that all the beneficent energies of nature are expressed with a profusion of beauty which is not necessary to their efficient operation, we accept the manifestation, are made happy in the consciousness of being beloved, and, constituted as we are in the image and likeness of God, express our instinctive gratitude in those fine human sympathies which impress the seal of truth on the primary idea of our creation.

But if our

> "little, nameless, unremembered acts
> Of kindness and of love"

are a reflection of the divine spirit within us, and if our capacity for enthusiasm, in the primary etymological use of the word — that is, our ability to be possessed or inspired by the spirit of a god, — is one more proof that we are created in the image of Deity, there is still another and more significant mark of relationship.

The human heart, in its best estate, forever yearns *to create*, to give in some form expression and life to those evasive ideals of loveliness which are planted within our consciousness by the aspects of nature. Our instinctive longing to embody these illusive phantoms of beauty is godlike. But the immortal songs which remain unsung, the exquisite idyls which gasp for words, the bewildering and restless imagery which seeks in vain the eternal repose of marble or of canvas, — while these reveal the affectionate and divine desires of humanity, they prove, also, how few there are to whom it is given to learn the great lesson of creation. When one arises among us who, like Pygmalion, makes no useless appeal to the Goddess of Beauty for the gift of life for his ideal, and who creates as he was created, we cherish him as a great interpreter of human love. We call him poet, composer, artist, and speak of him reverently as *Master*. We say that his lips have been wet with dews of Hybla; that, like the sage of Crotona, he has heard the music of the spheres; that he comes to us, another Numa, radiant and inspired from the kisses of Egeria.

Thus, as infinite Love begets infinite Beauty, so does infinite Beauty reflect into finite perceptions that image of its divine parentage, which the antique world worshiped under the personification of Astarte, Aphrodite, Venus, and recognized as the *great creative principle* lying at the root of all high Art.

There is a curious passage in Boehme, which relates how Satan, when asked the cause of the enmity of God and his own consequent downfall, replied, — "I wished to be an Artist." So, according to antique tradition, Prometheus manufactured a man and woman of clay, animated them with fire stolen from the chariot of the sun, and was punished for the crime of creation; Titans chained him to the rocks of the Indian Caucasus for thirty thousand years.

This Ideal, this Aphrodite of old mythologies, still reigns over the world of art, and every truly noble effort of the artist is saturated with her spirit, as with a religion. It is impossible for a true work of art to exist, unless this great creative principle of love is present in its inception, in its execution, in its detail. It must be pervaded with the warmth of human, passionate affection. The skill which we are so apt to worship is but the instrument in the hands of love. It is the means by which this humanity is transferred to the work, and there idealized in the forms of nature. Thus the test of art is in our own hearts. It is not something far away from us, throwing into our presence gleaming reflections from some supernal source of light and beauty; but it is very near to us, — so near that, like the other blessings which lie at our feet, we overlook it in our far-reaching searches after the imaginary good. We, poor underlings, have been taught in the school of sad experience the mortal agony of love without skill, the power of perception without the power of utterance. We know how dumb are the sweet melodies of our souls, how fleeting their opulent and dreamy pageantries. But we have not fully learned the utter emptiness and desolation of skill without love. We accept its sounding brass and tinkling cymbals for immortal harmonies. We look reverently upon its tortured marbles, its architectural affectations, its canvases stained with academic knowledge, as revelations of higher intelligence; unaware that if we go down to the quiet places of our own souls, we shall find there the universe reflected, like a microcosm, in the dark wellsprings, and that out of these wellsprings in the deep silence rises the beautiful ideal, Anadyomene, to compensate and comfort us for the vacancy of life. If we know ourselves, it is not to the dogmas of critics, the artificial rules of æsthetics, that we most wisely resort for judgments concerning works of art. Though technical externals, though conventionalities, sophisticated by the complications of modern life, and the address of manipulation naturally take possession of our senses and warp our opinions, there are depths of immortal truth within us, rarely sounded, indeed, but which, when dimly revealed in some rare moment of insight, seem to show that we possess a standard and a criterion far nobler than the schools can give us.

The broken statues and columns and traditions and fragmentary classics, which Greece has left us, are so still and tranquil to the eye and ear, that we search in vain for the Delphic wisdom they contain, till we find it echoed in the sympathetic depths of our souls, and repeated in the half-impalpable ideals there. It is in this favored land of art that we must seek for the earliest, the clearest, and the purest external expressions of these ideals, whose existence we but half suspect within us. It is not pleasant, perhaps, to think that we were nearly unconscious of some of the highest capacities of our humanity till we recognized their full expression in the ashes of a distant and dead pagan civilization; that we did not know ourselves till

> "The airy tongues that syllable men's names
> In pathless wildernesses"

uttered knowledge to us among the ruins of Hellas. It is good for us to lend a spiritual ear to these ancient whisperings, and hear nymph calling to nymph and faun to faun, as they caper merrily with the god Pan through the silence. It is good for us to listen to that "inextinguishable laughter" of the happy immortals of Olympus, ever mingling with all the voices of Nature and setting them to the still sweet music of humanity, — good, because so we are reminded how close we are to the outward world, and how all the developments of visible beauty are figurative expressions of our own humanity. Thus it is that the poetic truths of old religions exquisitely vindicate themselves; thus we find, even we moderns, with our downward eyes and our wrinkled brows, that we still worship at the mythological altars of childlike divinities; and when we can get away from the distracting Bedlam of steam-shrieks and machinery, we behold the secrets of our own hearts, the Lares and Penates of our own households, reflected in the "white ideals" on antique vases and medallions.

Abstract lines are the most concentrated expressions of human ideas, and as such are peculiarly sensitive to the critical tests of all theories of the Beautiful. Distinguished from the more usual and direct means by which artists express their inspirations and appeal to the sympathies of men; distinct from the common language of art, which contents itself with conveying merely local and individual ideas by imitations or translations of natural forms, abstract lines are recognized as the grand hieroglyphic symbolism of the aggregate of human thought, the artistic manifestations of the great human Cosmos. The natural world, passing through the mind of man, is immediately interpreted and humanized by his creative power, and assumes the colors, forms, and harmonies of Painting, Sculpture, and Music. But abstract lines as we find them in Architecture and in the ceramic arts are the independent developments of this creative power, coming directly from humanity itself, and obtaining from the outward world only the most distant motives of composition. Thus it is an inevitable deduction that Architecture is the most *human* of all arts, and its lines the most *human* of all lines.

"A thing of beauty is a joy forever;"

and the affectionate devotion with which this gift is received by finite intelligences from the hand of God is expressed in art, when its infinite depth *can* be so expressed at all, in a twofold language, — the one objective, the other subjective; the one recalling the immediate source of the emotion, and presenting it palpably to the senses, but more or less infused with the thought or mood of the artist; the other portraying rather the emotion than the cause of it, and, by an instinctive and universal symbolism, reflecting the character of the impression made by it upon the creative instinct and inner consciousness of mankind. Hence come those lines which æsthetic writers term "Lines of Beauty," suggesting rather than expressing meanings, — simple and elementary, but of far-reaching and various significance; animated with life and thought and musical motion, and yet still and serene and spiritual.

The mysteries of orb and cycle, with which old astrologers girded human life, and sought to define from celestial phenomena the horoscope of man, have been brought down to modern applications by learned philosophers and mathematicians, who have labored with ingenuity and skill to trace the interior relationships existing between the recondite revelations of their geometry, their wonderful laws of mathematical harmonies and unities, and those lines which are creations of art, as distinguished from imitations of nature, and which, because their various characteristics emanated from the inborn genius or spirit of the races of mankind, have become symbols of humanity. No well-organized intellect can fail to perceive that some sublime and immortal truth underlies these speculations. Undoubtedly, in the straight line, in the conic sections, in the innumerable composite curves of the mathematician, lie the germs of all these symbolic expressions. But the architect, whose lines of beauty vary continually with the emotions which produce them, who feels in his own human heart the irresistible impulse which gives to them their exquisite balance, their sensitive poise, their delicate significance and value, cannot allow that the spirit and motive of these lines are governed by the exact and rigid formulas of the mathematician to any greater extent or in any other manner than as the numbers of the poet are ruled by the grammar of his language. These formulas may be applied as a curious test to ascertain what strange sympathies there may be between such lines and the organic harmonies of nature and the universe; but these formulas do not enter into the soul of their creation any more than the limitations of counterpoint and rhythm laid their incubus on the lyre of Apollo. The porches where Callicrates, Hermogenes, and Callimachus walked were guarded by no such Cerberus as the disciples of Plato encountered at the entrance of the groves of the Academy, —

"Οὐδεὶς ἀγεωμέτρητος εἰσίτω,"
"Let no one ignorant of Geometry enter here;"

but the divine Aphrodite welcomed all mankind to the tender teachings of the lotus, the acanthus, the honeysuckle, and the seashell, and all the sweet relationships of nature to the heart of man.

The tendency of many modern writers to account for the lines of Greek mouldings and vases by considering them developments of strict scientific laws, results of mathematical proportions, and by endeavoring to restrict their freedom to the narrow limitations of geometric lines and curves, nicely adapted from the frigidities of Euclid, is dispiriting to the artist. "The melancholy days have come" for art when the meditative student finds his early footsteps loud among these dry, withered, and sapless leaves, instead of brushing away the dews by the fountains of perpetual youth.

Now it is capable of distinct proof that abstract lines of beauty, as in all their infinite variations of movement they are delicately sensitive to the moods or emotions of the artist, are results of the noblest creative capacities of mankind and expressions of the highest art.

The water-lily, or lotus, perpetually occurs in Oriental mythology as the sublime and hallowed symbol of the productive power in Nature, — the emblem of that great life-giving principle which the Hindu and the Egyptian and all early nations instinctively elevated to the highest and most cherished place in their Pantheons. Mr. W. H. Goodyear, in his recent essay on "The Ionic Capital and the Origin of the Anthemion," attempts very ingeniously to prove that from the drooping calyx leaves of this plant were derived some of the most characteristic decorative features of Greek art. But however this may be, Payne Knight many years ago attributed the adoption of this symbol to the fact that the lotus, instead of rejecting its seeds from the vessels where they are germinated, nourishes them in its bosom till they have become perfect plants, when, arrayed in all the irresistible panoply of grace and beauty, they spring forth, Minerva-like, float down the current, and take root wherever deposited. And so it was used by nearly all the early peoples to express the creative spirit which gives life and vegetation to matter. Lacshmi, the beautiful Hindu goddess of abundance, corresponding to the Venus Aphrodite of the Greeks, was called "the Lotus-born," as having ascended from the ocean in this flower. Here, again, is the inevitable intermingling of the eternal principles of Beauty, Love, and the Creative Power in that pure triune medallion image which the ancients so tenderly cherished and so exquisitely worshiped with vestal fires and continual sacrifices of art. Old father Nile, reflecting in his deep, mysterious breast the monstrous temples of Nubia and Pylæ, bears eloquent witness to the earnestness and sincerity of the old votive homage to Isis, "the Lotus-crowned" Venus of Egypt. For the symbolic water-lily, *recreated* by human art, blooms forever in bud or flower in the capitals of Karnac and Thebes, and wherever columns were reared and lintels laid throughout the length and breadth of the "Land of Bondage." It is the keynote of all that architecture and the basis of its decorative character. A brief examination into the principles of this new birth of the Lotus, of the monumental straightening and stiffening of its graceful and easy lines, will afford some insight into the strange processes of the human mind, when by divine instinct it creates out of the material beauties of nature a work of art.

It is well known that the religion of the old Egyptians led them to regard this life as a mere temporary incident, an unimportant phase of their progress toward that larger and grander state imaged to them with mysterious sublimity in the idea of Death or Eternity. In accordance with this belief, they expressed in their dwellings the sentiment of transitoriness and vicissitude, and in their tombs an immortality of calm repose. And so their houses have crumbled into dust ages ago, but their tombs are eternal. In all the relations of life the sentiment of death was present in some form or other. The hallowed mummies of their ancestors were the most sacred mortgages of their debts, and to redeem them speedily was a point of the highest honor. They had corpses at their feasts to remind them how transitory were the glory and happiness of the world, how eternal the tranquillity of death.

Now, how was this prevailing idea expressed in their art? They looked around them and saw that all organic life was full of movement and wavy lines; their much-loved Lotus undulated and bent playfully to the solemn flow of the great Nile; the Ibis fluttered with continual motion; their own bodies were full of ever-changing curves; and their whole visible existence was unsteady, like the waves of the sea. But when the temporary life was changed, and "this mortal put on immortality," when the mummy was swathed in its everlasting cerements for its long sleep, their eyes and souls were filled with the utter stillness and repose of its external aspects; its features became rigid and fixed, and were settled to an everlasting and immutable calm; the vibrating grace of its lines departed, and their ever-varying complexity became simplified, and assumed the straightness and stiffness of death. In this manner the straight line, the natural expression of eternal repose, in contradistinction to the wavy line, which represents the animal and vegetable movements of life, became the motive and spirit of their art. The anomaly of death in life was present in every development of the creative faculty, and no architectural or decorative feature could be so slight and unimportant as not to be thoroughly permeated with this sentiment. The tender and graceful lines of the Lotus became sublime and monumental under the religious loyalty of Egyptian chisels. Whether used in construction or decoration, these lines, so stiffened and formalized, seem to bring us into the presence of a strange phase of human life and thought, long since passed away from any other contact with living creatures. When applied to the rendering of the human figure, especially in the heroic forms of Egyptian art, — in the statues and portraits of kings, in the sphinxes, in the colossi of Memnon, and the royal tombs, — these solemn and fateful lines appear in effect to overwhelm the idea of human mutability and movement in the awful repose of immortality. It seems sufficiently evident that this process of formalizing the lines of the human figure in Egyptian art was not the result of a limitation of the powers of the artist to express movement and life, but rather of a consecration of his art to the service of a fundamental religious dogma.

"Solid-set,
And moulded in colossal calm,"

all the lines of this lost art thus recall the sentiment of endless repose, and even the necessary curves of its mouldings are dead with straightness. The love which produced

GREEK LINES.

1, 2, PALACE OF THOTHMES III., KARNAC.
3, PALACE OF AMENOPHIS III., THEBES.

these lines was not the passionate love which *we* understand and feel; they were not the result of a sensuous impulse; but the Egyptian artist seemed ever to be standing alone in the midst of a trackless and limitless desert, — around him earth and sky meeting with no kiss of affection, no palpitating embrace of mutual sympathy; he felt himself encircled by a calm and pitiless Destiny, the cold expression of a Fate from which he could not flee, and in himself the centre and soul of it all. Oppressed thus with a vast sense of spiritual loneliness, when he uttered the inspirations of art, the memories of playful palms and floating lilies and fluttering wings, though they came warm to the love of his heart, were attuned in the outward expression to the deep, solemn, prevailing monotone of his humanity. His love for the lotus and the ibis, more profound than the passion of the senses, dwelt serene in the bottom of his soul, and thence came forth transfigured and dedicated to the very noblest uses of life. This is the art of Egypt.

But among all the old nations which have perished with their gods, Greece appeals to our closest sympathies. She looks upon us with the smile of childhood, free, contented, and happy, with no ascetic self-denials to check her wild-flower growth, no stern religion to bind the liberty of her actions. All her external aspects are in harmony with the weakness and the strength of human nature. We recognize ourselves in her, and find all the characteristics of our own finite nature there developed into a theism so human, clothed with a personification so exquisite and poetical, that the Hellenic mythology seems still to live in our hearts a silent and shadowy religion without ceremonies or altars or sacrifices. The festive gods of the Iliad made man a deity to himself, and his soul the dwelling-place of ideal beauty. In this ideal they lived and moved and had their being, and came forth thence, bronze, marble, chryselephantine, a statuesque and naked humanity, chaste in uncomprehended sin and glorified in antique virtue. The beauty of this natural life, and the love of it, was the soul of the Greek ideal; and the nation continually cherished and cultivated and refined this ideal with impulses from groves of Arcadia, vales of Tempe, and flowery slopes of Attica, from the manliness of Olympic games and the loveliness of Spartan Helens. They cherished and cultivated and refined it, because here they set up their altars to known gods and worshiped attributes which they could understand. The ideal was their religion, and the art which came from it, the expression of their highest aspiration.

When the Apostle Paul read upon the Athenian altar the dedication "To the *Unknown* God," inscribed when the old mythology was becoming obsolete in the progress of philosophic doubt, he saw a manifestation of that Greek spirit, inquisitive, restless, but ever creative, which, three hundred years afterwards at Byzantium, made this people the first interpreters of Christianity in a new art. Heine, in his "Reisebilder," tells the strange story of the Homeric feast of the immortal gods upon Olympus, interrupted by a pale Jew, crowned with thorns, dripping with blood, and bearing a heavy cross of wood, which he threw with a thunderous crash upon the high table, and immediately the golden goblets were overturned and the pagan pageant faded from the sight forever; with it faded the Greek ideal, to be dimly recalled, as we shall presently see, after many centuries of complete slumber, in the rediscovery of the principle of Greek Lines.

Lines of beauty, produced in such a soil, were not, as might at first be supposed, tropic growths of wanton and luxurious curves, wild, spontaneous utterances of superabundant life. The finely studied perception of the Greek artist admitted no unimaginative, pre-Raphaelite imitations of what Nature was ever giving him with a liberal hand in the whorls of shells, the veins of leaves, the life of flames, the convolutions of serpents, the curly tresses of woman, the lazy grace of clouds, the easy sway of tendrils, flowers, and human motion. He was no literal interpreter of her whispered secrets. But the grace of his art was a *deliberate grace*,— a grace of thought and study. His lines were *creations*, and not *instincts* or *imitations*. It was his religion so to nurture and educate his sensitiveness to beauty and his power to love and create it, that his works of art should be deeds of worship and expressions of a godlike humanity. Unlike the Egyptian's, there was nothing in *his* creed to check the sweet excess of Life, and no grim shadow, "feared of man," scared him in his walks, or preached to him sermons of mortality in the stones and violets of the wayside. Life was hallowed and dear to him for its own sake. He saw it was lovable, and he made it the theme of his noblest poems, his subtlest philosophies, and his highest art. Hence the infinite joy and endless laughter on Olympus, the day-long feasting of the gods, the silver stir of strings in the hollow shell of the exquisite Phœbus, "the soft song of the muse with voices sweetly replying."

I believe that all true lines of grace and beauty, though they are necessarily infinite in variety and meaning, are capable of division into three distinct classes, according to the respective spirit or genius of the three civilizations out of which they grew, and that each class, as it expresses a unity of great significance in the history of the human race, is capable of concentrating its distinguishing characteristics in one representative line, which may stand as the symbol, the gesture, as it were, of an era in this history. If this is possible, these three symbols would represent not only the arts of three great eras, but the essential quality of their life and thought, of which these arts are themselves the symbol and the record. These lines, from the nature of the case, are not precise or exact, like a formula of mathematics, to which the neophyte can refer for deductions of grace to suit any premises or conditions. This, of course, is contrary to the spirit of beautiful design; and the ingenious Hay, who maintained that his "composite ellipse" is capable of universal application in the arts of ornamental composition, and that by its use any desirable lines in mouldings or vases can be mechanically produced, especially Greek lines, fell into the grave error of endeavoring to materialize and fix that *animula vagula, blandula,* that coy and evasive spirit of art, which is its peculiar characteristic, and gives to its works inspiration, harmony, and poetic sentiment. Ideal beauty can be hatched from no geometrical eggs. Of these lines one expresses the most subtle grace yet conceived by the mind or executed by the hand of man. This line pretends to be merely a type of that large language of forms, with which the most refined intellects of antiquity uttered their joyful worship of Aphrodite in Greek art.

The three great distinctive eras of art, in a purely psychological sense, were the Egyptian, the Grecian, and the Romanesque, — including in the latter term both Roman art itself and all subsequent art, whether derived directly or indirectly from Rome, as the Byzantine, the Mahometan, the Mediæval, and the Renaissance. Selecting the most characteristic works to which these great eras respectively gave birth, it is not difficult by comparison to ascertain the master-spirit, or type, to which each of these three families may be reduced. If we place these types side by side, the result will be as in the diagram,

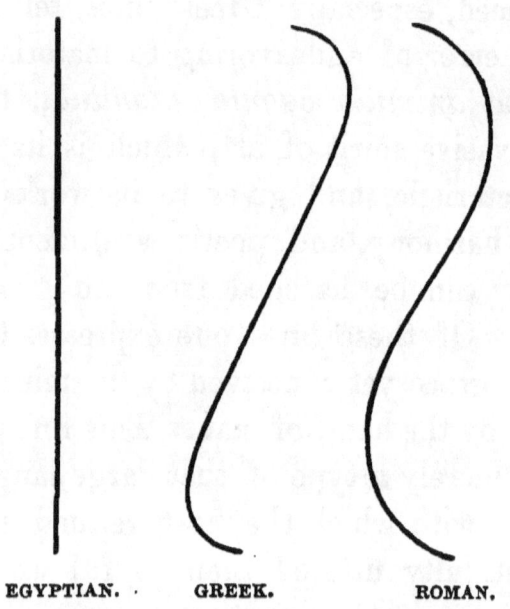

EGYPTIAN.   GREEK.   ROMAN.

presenting to the eye, at one view, the concentration of three civilizations, DESTINY, LOVE, and LIFE; — destiny, finding utterance in the stern and inflexible simplicity of the tombs and obelisks of Egypt; love, expressing itself in the statuesque and thoughtful grace of Grecian temples, statues, and urns; life, in the sensuous and impulsive change, evident in all the developments of art, since Greece became Achaia, a province of the Roman Empire. In the central type, as exhibited in the diagram, we see a happy mingling of the essential qualities of the other two. The genius of Greece, on the one hand, tempers the rigid repose of Egypt with the passion of life, and, on the other, restrains the passion of life with a spirit of intellectual tranquillity. Humanly speaking, it seems to stand for a condition of perfect art, representing the highest development of creation by man. At one point in the history of the race the secret of perpetual youth seemed to have been disclosed, at once grave with memory and jocund with hope. This intermediate line is the essence of the age of Pericles; and in it "the capable eye" may discover the pose of the Cnidian Venus of Praxiteles, of the Jupiter Olympus of Phidias, and the other lost wonders of ancient chisels, and, more directly, the tender severity of Doric capitals and the secret grace of the shafts of the Parthenon.

According to Pliny, Apelles, when he visited the great painter Protogenes, at Rhodes, and found him not at home, inscribed a *line* upon a board, assuring the slave that this line would signify to the master who had been to see him. Whatever the line was, Protogenes, we hear, recognized in it the hand of the greatest limner of Greece. It was the signature of that Ideal, known to the antique world by its wider developments in the famous pictures of the Venus Anadyomene, and Alexander with the Thunderbolt, hung in the temple of Diana at Ephesus.

The gravity with which this apparently trifling anecdote is given us from antiquity evidently proves that it was one of the household tales of old Greece. It did not seem absurd in those times, when art was recognized as a great unity, an elaborate and copious language, founded on the simplest elements of life, and, in its grandest and widest flowings, bearing ever in its bosom, like a great river, the memory of the little weeping Naiad far up among the mountains with her "impoverished urn."

Thus every great national art, growing up naturally out of the necessities of an earnest people, expressing the grand motives of their life, as that of the Greeks and the Egyptians and the mediæval nations of Europe, is founded on the simplest laws. So long as these laws are obeyed without affectations or sophistications, art is good and true; so long as it remembers the purity and earnestness of its childhood, the strength that is ordained out of the mouth of babes is present in all its expressions; but when it spreads itself abroad in the fens and marshes of humanity, when it becomes reminiscent and imitative, when it is governed by archæology and pretends to be what it is not, it loses the purity of its aim, the singleness and unity of its action, — it becomes stagnant, and sleeps in the death of idleness.

When one studies Greek art, the whole motive of it seems so childlike and so simple that the impulse to seek for that little Naiad which is the fountain and source of it all is irresistible. I have endeavored, in the representative line which I have traced, to embody the characteristic grace of the delicate and elegant lines in the profiles of Greek mouldings and vases, and to present a type of the restrained and carefully balanced rhythmic movement, the aristocratic poise, which distinguish the composition of Greek ornament. It is impossible to produce this line with a wanton flourish of the pencil, as I have done in that easy and thoughtless curve which Hogarth, in his quaint "Analysis of Beauty," assumed as the line of true grace; nor yet are its motions governed by any cold mathematical laws. In it is the earnest and deliberate labor of love.

There are thought and tenderness in every instant of it; but this thought is grave and almost solemn, and this tenderness is chastened and purified by wise reserve. Measure it by time, and you will find it no momentary delight, no voluptuous excess which comes and goes in a breath; but there is a whole cycle of deep human feeling in it. It is the serene joy of a nation, and not the passionate impulse of a man. Observe, from beginning to end, its intention is to give expression by the serpentine line to that sentiment of beautiful life which was the worship of the Greeks; but they did not toss it off, like a wine-cup at a feast. They prolonged it through all the varied emotions of a lifetime with exquisite art, making it the path of their education in childhood and of their wider experience as men. All the impulses of humanity they bent to a kindly parallelism with it. This is that famous principle of Variety in Unity which St. Augustine and hosts of other philosophers considered the true ideal of Beauty. Start with this line from the top upon its journeying: look at the hesitation of it ere it launches into action; how it cherishes its resources, and gathers up its strength! — with a confidence in its beautiful Destiny, and yet with a chaste shrinking from the full enjoyment of it, how inevitably, but how purely it yields itself to the sudden curve! It does not embrace this curve with a sensuous sweep, nor does it, like Sappho, throw itself with quick passion into the tide. It enters with maidenly and dignified reserve into a new life which, as you glance at it, seems almost ascetic, and reminds you of the rigid fatalism of Egypt. Its grace is almost strangled, as those other serpents were in the grasp of the child Hercules. But if you watch it attentively, you will find it ever changing, though with subtlest refinement, ever human, and true to the great laws of emotion. There is no straight line here, no death in life, but the composure of " maiden meditation " moving with serious pleasure along the grooves of happy change. It certainly does not seem to be merely fanciful to attribute the character of this line to a habit of chastity in thought and of elegant reserve in expression. Follow it still farther, and you will find it grateful to the sight, neither fatiguing with excess of monotony nor cloying the appetite with change. And when the round hour is full and the end comes, this end is met by a Fate, which does not clip with the shears of Atropos and leave an aching void, but fulfills itself in gentleness and peace. The line bends quietly and unconsciously towards the beautiful consummation, and then dies, because its work is done.

When we examine the innumerable lines of Greek architecture, we find that the artists never for an instant lost sight of this ideal. Its deliberate and carefully studied grace was everywhere present, and mingled not only with such grand and heroic lines as those of the sloping pediments, the long-drawn entablatures, and massive stylobates of the Parthenon and Theseion, bending them into curves so subtly modulated that our coarse perceptions did not perceive the variations from the dead straight lines till the careful admeasurements of Penrose and Cockerel and their *confrères* of France assured us of the fact, still further confirmed by the skillful photographs of Stillman; not only did it make these enormous harp-strings vibrate with deep human soul-music, but there is not an abstract line in moulding, column, or vase, belonging to old Greece or the islands of the Ægean or Ionia or the Greek colonies of Italy, which does not have the same deep meaning, the same statuesque and thoughtful pose. Besides, I very much doubt if the same line, in all its parts and proportions, is ever repeated twice, — certainly not with any emphasis; and this is following out the great law of our existence, which varies the emotion infinitely with the occasion which produced it. Let us suppose, for example, that a moulding was needed to crown a column with fitting glory and grace. Now the capital of a column may fairly be called the throne of ideal expression; it is the *cour d'honneur* of art. The architect in this emergency did not set himself at "the antique," and seek for authorities, and reproduce and copy; for he desired not to present with pedantic accuracy and historical respect some effect of beautiful design which he had seen in the capitals of Assyria, or Persia, or Egypt, and which had been developed from some condition of structure, or use, or habit of mind different from his own, but to devise a feature which physically and æsthetically should answer all the requirements of its peculiar position, a feature whose lines should have essential relationships with the other lines around it, those of shaft, architrave, frieze, and cornice, — should swell its fitting melody into the great fugue. And so, between the summit of the long shaft and that square block, the *abacus*, on which reposes the dead weight of the lintel of Greece, the

Doric *echinus* was fashioned, crowning the serene Atlas-labor of the column with exquisite completion, and uniting the upright and horizontal masses of the order with a marriage ring, whose beauty is its perfect fitness. The profile of this moulding may be rudely likened to the upper and middle parts of the line assumed as the representative of the Greek ideal. But it varied ever with the exigency of circumstances. Over the short and solid shafts of Pæstum it assumed the shape of a crushed cushion; that of the Acropolis at Selinus (shown in diagram at A) had a similar character. The conditions in these early examples of the style seemed to the artist, as the shaft had an abundance of reserved strength, to require at this point a moulding of softened grace, while the Parthenon (shown at B) and the temple of Ceres at Eleusis (shown at C) exhibited in their *echini* more energetic profiles, as if their service of support could not be performed satisfactorily to the eye without an expression of effort, somewhat like that of a muscle in strong tension. The comparatively slender grace of the shafts in these two latter and later examples — a grace emphasized by the almost imperceptible vertical curves with which these shafts diminish, in the

GREEK LINES. 33

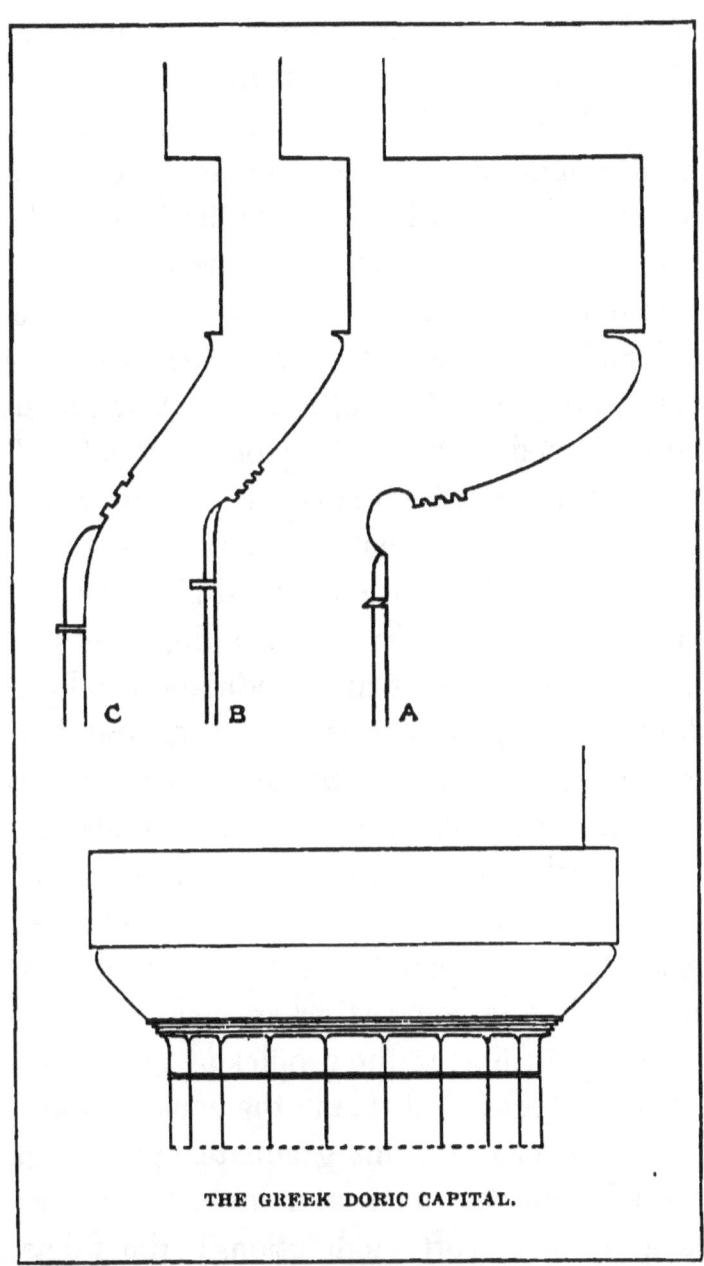

THE GREEK DORIC CAPITAL.

entasis, from base to summit — demanded at the bearing point a demonstration of power to justify their delicacy of proportion. The especial profile of the *echinus* in these cases seemed to be preordained from the base of the shaft, like a flower from the root. It was created as with "the Dorian mood of soft recorders." Between these two extremes there is an infinity of change, everywhere modified and governed by "the study of imagination."

The same characteristics of nervous grace and severe intellectual restraint are found wherever the Greek architect put his hand and his heart to work. Every moulding bears an impress of refinement, and modulates the light which falls upon it with exquisite and harmonious gradations of shade. The sun, as it touches the mouldings of the entablature, makes visible music there, as if it were the harp of Memnon, — presenting a series of horizontal and parallel shade and shadow lines, delicately contrasting in melodious succession, according to the profiles of the mouldings which produced them; the cymas, ovolos, and beads, creating fine graduated pencilings of shade, whose upper and lower boundaries die away with soft modulations; the fillets, throwing sharp and thin shadows along the

GREEK LINES.

illuminated marble surfaces; the overhanging corona making a broad, defined shadow, broken by reflected lights playing on the supporting bed-mouldings. All the phenomena of reflected lights, half lights, and broken lights are brought in and attuned to the great dædal melody of the edifice. The antiquities of Attica afford nothing frivolous or capricious or merely fanciful, no playful extravagances or wanton meanderings of line; nor did they ever submit to the tyranny of conventional formulas; but, ever loyal to the purity of a high ideal, they give us, even in their ruins, a wonderful and very evident unity of expression, pervading and governing every possible mood and manner of thought.

No phase of art that ever existed gives us a line so very human and simple in itself as this Greek type, and so pliable to all the uses of monumental language. If this type were a mere mathematical type, its applicability to the expression of human emotions would be limited to a formalism absolutely fatal to the freedom of thought in art. But because it has its birth in love, as distinguished from pedantry, from scholastic or archæological conformity; because it indicates a refined appreciation of all the movements of life and all the utterances of creation; because it is the humanized essence of these motions and developments, it becomes a prolific unity, containing within itself the germs of a new world of ever new delight.

When this type in Greek art was brought to bear on the interpretation of natural forms into architectural language, we shall curiously discover that the creative instinct of the artist and his reverence for the integrity of his ideal were so great, that he not only subjected these forms to a rigid subservience to the abstract line till nature was nearly lost in art, but the adoption of these forms under any circumstances was limited to some two or three of the most ordinary vegetable productions of Greece and to one seashell.[1] This wise reserve and self-restraint present to us the best proof of the fastidious purity of his thought. Nature poured out at the feet of the Greek artist a most plenteous offering, and the lap of Flora overflowed for him with tempting garlands of beauty; but he did not gather these up with any greedy and indiscriminate hand, he did not intoxicate himself at the harvest of the vineyard. If he adorned his festivals

[1] Mr. Goodyear, as we have seen, argues that they were all derived from one, the Egyptian lotus.

profusely with garlands and wreaths of flowers, he decorated his works of art with serious and reverential reserve, contenting himself with two or three simple types, which he varied with an invention at once chaste and elegant. Intent upon the nobler aim of creating a pure work of art, he considered what expression or character of line or movement was required in the surface which he was to decorate, whether frieze, soffite, necking, border, moulding, capital, or vase, and taking two or three of the most familiar products of the soil, he studied their principles of growth, their creative sentiment, and adjusted them to his decorative purpose, in fact re-created them. The design was not bent to accommodate them, but they were translated and lifted up into the sphere of art. It is not impossible that the Greek artist, in this process of secondary creation, may have been influenced more or less by certain crude traditionary forms, which had reached him from the Nile and the Euphrates; for the artistic instinct quickly appreciates and absorbs any motive of design which may pass under his observation. But the Greek work was developed so far beyond all these precedents that, if possibly his initiative may be questioned, his supreme creative power cannot for a moment be disputed.

If we examine the magnificent Ionic capitals of the porch of Athene Polias in the Erechtheum, we may see that the Greek could be profuse without being redundant.

GREEK IONIC.

Whether the main idea of this lovely and elaborate creation came from some archaic type of Ionia or Persia or Assyria, or whether it was drawn from the convolutions of a seashell, he made "old things new" with his perfect art, and remodeled whatever they may have given him of suggestion into a composition of exquisite harmony and order. Examine the volute: this is the nearest approach to a mathematical result that can be found in Greek architecture; yet this very approximation is one of the greatest triumphs of art. No geometrical rule has been discovered which can exactly produce the spirals of the Erechtheum, nor can they be found in shells. In avoiding the exuberance of the latter and the rigid formalism of the former, a work of human thought and love has been evolved. Follow one of these volutes with your eye from its centre outwards, taking all its congeries of lines into companionship; you find your sympathies at once strangely engaged. There is an intoxication in the gradual and melodious expansion of these curves. They seem to be full of destiny, bearing you along, as upon an inevitable tide, towards some larger sphere of action. Ere you have grown weary with the monotony of the spiral, you find that the system of lines which compose it gradually leave their obedience to the centrifugal forces of the volutes, and, assuming new relationships of parts, sweep gracefully across the summit of the shaft, and become presently involved in the reversed motion of the other volute, at whose centre Ariadne seems to stand, gathering together all the clues of this labyrinth of beauty. This may seem fanciful to one who regards these things as matters of convention. But inasmuch as, to the studious eye of affection, they suggest human action and human sympathies, this perhaps may be accepted as a proof that they had their birth in some corresponding affection. It is the inanimate body of Geometry made spiritual and living by the creative power in the human heart. And when a later generation reduced the Ionic volutes to rule, and endeavored to inscribe them with the gyrations of the compass, they have no further interest for us, save as a mathematical problem with an unknown value equal to a mysterious symbol x, in which the soul takes no comfort. When we reasonably regard the printed words of an author, we not only behold an ingenious collection of alphabetical symbols, but are placed by them in direct contact with the mind which brought them together, and, for the moment, our train of thought so entirely coincides with that of the writer, that, though perhaps he died centuries ago, he may be said to live again in us. The capitals of the Erechtheum have the same renaissance in every heart capable of being touched by a work of art, which, though conventional in form, is inspired by the breath of imagination. The spirit of the ancient artist lives again in us for a happy moment, and the beautiful thing has a new existence in our affections. Studying it, we become artists and poets ere we are aware. The alphabet becomes a living soul.

Under the volutes of this capital, and belting the top of the shaft, is a broad band of ornamentation so happy and effectual in its uses, and so pure and perfect in its details, that a careful examination of it will, perhaps, afford us some knowledge of that spiritual essence in the antique ideal out of which arose the silent and motionless beauty of Greek marbles.

Here are brought together the *sentiments* of certain vegetable productions of Greece, but sentiments so entirely subordinated to the flexure of the abstract line that their natural significance is almost lost in a new and more human meaning. Here is the honeysuckle, the wildest, the most elastic and undulating of plants, under the severe discipline of order and artistic symmetry assuming a strict and chaste propriety, a formal elegance, which render it at once monumental and dignified. The harmonious succession and repetition of parts, the graceful contrasts of curves and their strict poise and balance, their unity in variety, their entire subjection to æsthetic laws, — these qualities combine in the creation of one of the most beautiful, most useful, and most effective devices of decoration ever applied to a work of art. It is called the Ionic *Anthemion,* and suggests in its composition all the creative powers of Greece. Its value is not alone in the sensuous gratification of the eye by its elegant symmetries and its formal repetitions, as with the Arabesque tangles of the Alhambra, but it is more especially in its intellectual expression, its fastidious and highbred poise, — the evidence there is in it of thoughtfulness and judgment and deliberate care in the adjustment of natural forms and movements to decorative uses. The inventor studied not alone the plant as it grew by nature, but as it might grow in the garden of art when informed with a spirit of humanity; and ere he made his interpretation he considered perhaps how, in mythological traditions, each flower once bore a human shape, and how Daphne and Syrinx, Narcissus and Philemon, and those other idyllic beings, were eased of the stress of human emotions by becoming Laurels and Reeds and Daffodils and sturdy Oaks, and how human nature was thus diffused through all created things and was epigrammatically expressed in them.

> "And he, with many feelings, many thoughts,
> Made up a meditative joy, and found
> Religious meanings in the forms of nature."

Like Faustus, he was permitted to look into her deep bosom, as into the bosom of a friend; to find his brothers in the still wood, in the air, and in the water; to see himself and the mysterious wonders of his own breast in the movements of the elements. And so he took Nature as a figurative exponent of humanity, and extracted the symbolic truths from her productions and used them nobly in his art.

Garbett, an English æsthetical writer, assures us that the *Anthemion* bears not the slightest resemblance to the honeysuckle or any other plant, "being no representation of anything in nature, but simply the necessary result of the complete and systematic attempt to combine unity and variety by the principle of *gradation*." But here he speaks like a geometer, and not like an artist. He seeks rather for the resemblance of form than the resemblance of spirit, and, failing to realize the object of his search, he endeavors to find a cause for this exquisite effect in pure reason. With equal perversity, Poe endeavored to persuade the public that his "Raven" was the result of mere æsthetical deductions!

If, in brief, we would know the secret of the Greek ideal we must first learn what virtues are hidden in simplicity and truth, and love with the wisdom and chastity of old Hellenic passion. We must sacrifice, not only undisciplined invention, but cold academic formulas, the precepts of pedants, the dogmas of precise archæologists, of architectural geometers and botanists, whose specious superficialities are embodied in the errors of modern art, — we must sacrifice these at the shrine of the true Aphrodite; else the modern Procrustes will continue to stretch and torture Greek lines on geometrical beds, and the æsthetic Pharisees around us will still crucify the Greek ideal.

II.

> "As when a ship, by skillful steersman wrought
> Nigh river's mouth or foreland, where the wind
> Veers oft, as oft so steers, and shifts her sail, —
> So varied he, and of his tortuous train
> Curl'd many a wanton wreath in sight of Eve,
> To lure her eye."

And Eve, alas! yielded to the blandishments of the wily serpent, as we moderns, in our art, have yielded to the Roman ideal of beauty, symbolized in the easy flowing and licentious curve which we have ventured to call an expression of Life, as contrasted with those severer lines which were created by Love in the ideal of Greece, and by Destiny in that of Egypt. This is the line made familiar to us by the thoughtless flourish of Hogarth's pencil, and, in a far wider sense, it is the line which has controlled all the demonstrations of modern art since the time of the Roman Empire, because this art has been rather imitative than creative and has held a too faithful mirror up to nature. At all events it has been content to let the purer ideal remain petrified in the marbles of Greece for nearly nineteen centuries.

I have endeavored to show how the Greek ideal may be concentrated in a certain abstract line, not only of sensuous, but of intellectual beauty. I have endeavored to prove how this line, the gesture of Attic eloquence, expresses the civilization of Pericles and Plato, of Euripedes and Apelles. It is now proposed briefly to relate how this line was lost, when the politeness and philosophy, the literature and the art of Greece were chained to the triumphal cars of Roman conquerors, — and how it seems to have been found again in our own day, after slumbering so long in ruined temples, broken statues, and cinerary urns.

The scholar who studies the history of Greek art has a melancholy interest, like a surgeon, in watching its slow but inevitable atrophy under the incubus of Rome. The wise but childlike serenity and cheerfulness of soul, so tenderly pictured in the white stones from the quarries of Pentelicus, had, it is true, a certain sickly, exoteric life in Magna Græcia, as Pompeii and Herculaneum have proved to us. But the brutal manhood of Rome overshadowed and tainted the gentle exotic like a Upas-tree. The imported Greek, indeed, was permitted by his Roman master to work with a degree of freedom wherever this freedom did not interfere with that expression of insolent prodigality, of vehement and sensuous splendor, which the Empire desired to present in all its public works as a matter of political policy. In such cases, as in the decoration of halls and chambers, this exotic grew up into a dim resemblance of its ancient purity under other skies. It had, I think, an elegiac plaintiveness in it, like a song of old liberty sung in captivity. It was not stimulated by the constant pressure of public sympathy and interest, which under more congenial skies made possible the perfect development of the Greek ideal. In its more monumental developments, under these new influences, the true line of beauty became gradually vulgarized, and, by degrees, less intellectual and pure, till its spirit of fine and elegant reserve was quite lost in a coarse splendor. It must be admitted, however, that not a little of the old refinement may be found in the lamps and candelabra and vases and *bijouterie* which we have exhumed from the ashes of Vesuvius, and rescued from the deep detritus between the hills of Rome.

But, considering what has been left to us of the magnificent architecture of the Romans, we cannot but be impressed by the fact that, in decorating their enormous fabrics with the Greek orders, they used them as barbaric conquerors used the spoils of war. The delicate Greek lines were at once bent to answer uses of ostentation and not of art. In fact, a comparison of the architectural details of a Greek with those of a Roman monument reveals the contrasts between the genius of the two races more clearly than is shown in any other visible demonstration. To narrow a broad subject down to an illustration, let us look at a single feature, the *cyma reversa,* or the Lesbian *cyma,* as it was understood in Greece and Rome. This is a supporting moulding of very frequent occurrence in classic entablatures, a curved surface with a double flexure. Perhaps the type of Greek lines, as represented in the previous paper on this subject, may be safely accepted as a fair example of the Greek interpretation of this feature. The Romans, on the other hand, not being able to understand and appreciate the delicacy and deep propriety of this line, seized their compasses, and, without thought or love, mechanically produced a gross likeness to it by the union of two quarter-circles, thus : —

GREEK LINES.

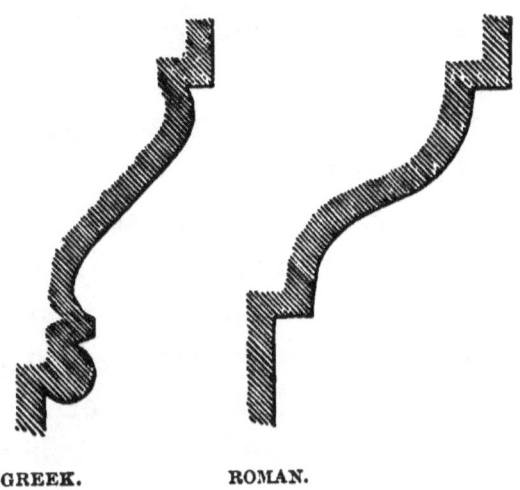

GREEK. ROMAN.

Look upon this picture, and on this! — the one, refined, delicate, fastidious, sensitive as a facial outline, and varying its character according to its especial function in the group of mouldings to which it belongs; the other, thoughtless, mathematical, sensuous, reappearing ever with a stolid monotony. Roman architecture was ever bold, ingenious, and altogether admirable in its structure and in its adaptation to practical and political uses; and in respect to the decoration of its architecture, it was magnificently redundant and prodigal. But though it respected the art of Greece, from which it borrowed all its finer qualities of detail, it accepted the letter for the spirit, and converted Greek mouldings into inflexible formulas. But even this *letter*, when they transcribed it, writhed and was choked beneath hands which knew better the iron cæstus of the gladiator than the subtle and spiritual touch of the artist.

We can have no stronger and more convincing proof that architecture is the truest record of the various phases of civilization than we find in this. Subjected to the imperial power and prodigal wealth of Rome, the Greek slaves were ready to teach their masters the literature, the philosophy, and art of their native country, — to conquer their conquerors with the refinements of a higher civilization. But Rome had her own ideas to enunciate; and so possessed was she by the impulse to give form to these ideas, to her ostentation, her pride, her sensuous magnificence, that she could not pause to learn calm and serious lessons from the Greeks who crowded her forums, but, seizing their fair sanctuaries, she stretched them out to fit her standard. She took the pure Greek orders to embellish her endless arcades; she piled these orders one above the other; she bent them around her gigantic circuses, she decorated every moulding of them lavishly with leaves, ovolos, or strings of pearls, and loaded every frieze with pompous inscriptions or with crowded undulations of wreathed foliage; she constructed their shafts with precious Oriental marbles and crowned them with capitals of gilded bronze; she applied them as decorations to her triumphal arches and to the immense vaulted halls in her baths and basilicas; till at last they had become acclimated and lost all their peculiar refinement, all their intellectual and dignified humanity. Every moulding, every capital, every detail, was formulated so as to be capable of ready reproduction. Out of the suggestions contained in the little delicate capitals of the choragic monument of Lysicrates and the Temple of the Winds at Athens, she developed that peculiarly Roman institution, the sumptuous Corinthian order; and added to it a variation if possible still more gorgeous, known as the Composite. These she established as official standards of her luxury and power, and with them she overawed the barbarians in the remotest colonies of her empire.

The Romans had neither time nor inclination to bestow any love or thought on the expressiveness and especial significance of subordinate parts. But out of the suggestions and reminiscences of Greek lines they made a rigid and inflexible grammar of their own, —

a grammar to suit the mailed clang of Roman speech, which, in its cruel martial strength, sought no refinements, no delicate inflections from a distant Acropolis. The result was the insolent splendor of the Empire. How utterly the still Greek ideal was forgotten in this noisy ostentation, how entirely the chaste

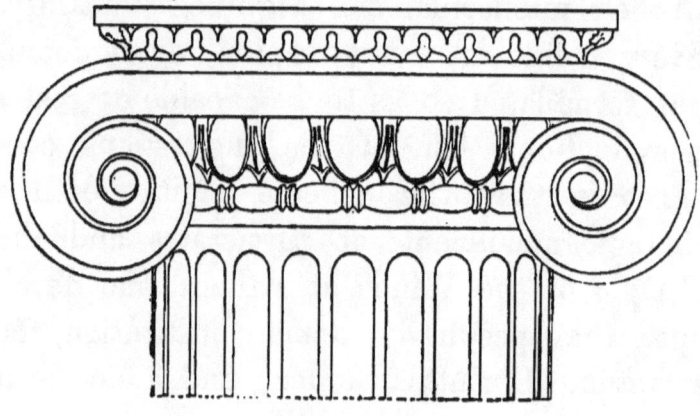

ROMAN IONIC.

spirituality of the Greek line was lost in the round and lusty curves which are the inevitable footprints of sensual life, scarcely needs further amplification.

I have referred to the Ionic capital of the Erechtheum as containing a microcosm of Attic art, as presenting a fair epitome of the thought and love which Hellenic artists offered in the worship of their gods. Turn now to the Roman Ionic, as developed in any one of the most familiar examples of it, in the Temple of Concord, near the Via Sacra, in the Theatre of Marcellus, or the Colosseum, and we shall see the results of organizing art into a political system. The capital, so formulated, has become mechanical and pattern-like. The grace of its freedom, its intellectual reserve, the secret humanity which thrilled through all its lines, the divine art in the rhythmical harmony of its movement, — all these are gone. Quality has yielded to quantity, and nothing is left save those external characteristics which he who runs may read, and he who pauses to study finds cold, vacant, and unsatisfactory. What the Ionic capital of Rome wants, and what all Roman art wants, is *the inward life*, the living soul, which gives a peculiar expressiveness to every individual work, and raises it infinitely above the dangerous academic formalism of the schools.

The practical point aimed at by the thoughtful and conscientious Greek master, aware that every fellow-citizen was interested in his work, and would not fail to subject it to critical analysis, was that his entablatures should be composed of mouldings so delicately adjusted that each one should play its properly subordinated part in the general composition, and, in the clear atmosphere of Greece, should make such graduated effects of shadows and shades in contrast with light as, in combination with those produced by the other members, should create a result of fine but complicated harmony. In this way he played delicately with sunlight, as a musician plays with sounds. His horizontal lines of light and dark were broken, where necessary for the sake of punctuation, by the regular interruptions of mutules, triglyphs, metopes, and dentils, each being scaled so as to have its due part in the whole design. He refined his lines and mouldings to such a point of sensitive delicacy that the least variation in them would produce to his fastidious eye an effect of discord. The profiles of his mouldings became as sensitive as the lines of the brows, the nose, or the mouth on the face of Apollo. These lines could not be altered by a hair's breadth without affecting the sentiment of the god-like ideal. Greek structure became in fact a highly organized unity, in which there was no mere perfunctory or conventional element, each detail being essential to the whole.

The Romans were prompt to imitate, but they could not assimilate an organism so fine. Their intensely practical and energetic spirit was too impatient to worship at a shrine requiring a service of constancy and love. The expression of unity was lost. Their structure was one thing and their decoration was another, and the decorative element was converted into a system of convention, as shown by the exact precepts of their own Vitruvius. This arrangement enabled the conqueror, without waste of time in that long contemplative stillness out of which alone the beauty of the true ideal arises, out of which alone man can create like a god, to avail himself at once of the Greek orders, not as a sensitive and delicate means of fine æsthetic expression, but as a mechanical language of contrasts of form to be used according to the exigencies of duty. The service of Greek art was perfect freedom; enslaved at Rome, it became academic. Thus systematized, it is true, it awes us by the superb redundancy and sumptuousness of its use in the temples and forums reared by that omnipresent power from Britannia to Baalbec. But the art which is systematized is degraded. Emerson somewhere remarks that man descends to meet his fellows, — meaning, I suppose, that he has to sacrifice some of the higher instincts of his individuality when he desires to become social, and to meet his fellows on that low level of society which, made up as it is of many individualities, has none of those secret aspirations which arise out of his own isolation. Society is a systematic aggregation for the benefit of the multitude, but great men lift themselves above it into a purer atmosphere. As Longfellow says, "They rise like towers in the city of God." So with art, — when we systematize it we degrade it, and deprive it of life and the power of progression. A singular proof of this is found in the fact that the Roman orders never have anything in them reserved from the common ken. They are reduced to rule, and therefore superficial. They were not based upon principles, like the Greek orders, and consequently the necessary modifications to which they, like all other formulas of fine art, must be subjected in use have been modifications of caprice rather than of structural development. Indeed, the Greek orders, in the hands of the Romans, gradually assumed characteristic forms which they never would have taken upon themselves if they had continued to be expressions of building as a practical art.

Virtually, the Roman orders died in the first century of the Christian era. We all know how, when the authority of the pagan schools was gone and the stern Vitruvian laws had become lost in the mists of antiquity, these orders gradually fell from their strict allegiance, and art was made free again by the enfranchised Greeks of Byzantium, under an entirely new religious impulse from Palestine. Structure and the decoration of structure were once more united, but in a fresh, progressive system, full of animation, power, and color, looking not backward to the contemplative ideal of the pagan Greeks, but forward eagerly to a triumphant demonstration of the new faith by art. The new spirit was rude and vigorous, but sincere and aspiring. Its typical line was one of robust and vigorous life, exuberant, because of the final safe establishment of Christianity after the persecutions of Aurelian and Diocletian. In the Romanesque of the West this line did not vary essentially in its character; but it developed and enlarged in its expressions, and, with the establishment and organization of the faith, it gradually assumed a characteristic of serious purpose. But, in after Gothic times, the spirit of the forgotten Aphrodite, ideal beauty, sometimes seemed to lurk furtively in the image of the Virgin Mary, and to inspire the cathedral builders with somewhat of the old creative impulse of love. But the workings of this impulse are singularly contrasted in the productions of the pagan Greek and the mediæval artists. Nature, we have seen, offered to the former mysterious and oracular Sibylline leaves, profoundly significant of an indwelling humanity diffused through all her woods and fields and mountains, all her fountains, streams, and seas. Those meditative creators sat at her feet, earnest disciples, but gathering rather the spirit and motive of her gifts than the gifts themselves, making an ideal and worshiping it as a deity. But for the cathedral builder, Dryads and Hamadryads, Oreads, Fauns, and Naiads did not exist, — the Oak of Dodona uttered no oracles.

> "A primrose by the river's brim
> A yellow primrose was to him,
> And it was nothing more."

To him nature was an open book, from which he continually quoted with a loving freedom, not to illustrate his own deep relationships with her, but to give greater glory to that vast Power which stood behind her beautiful text and was revealed to him in the new religion from Palestine. He loved fruits and flowers and leaves because they were manifestations of the love of God; and he used them in his art, not as motives out of which to create abstract forms, out of which to develop an ideal humanity, but to show his intense appreciation of the Divine love which gave them. Had he been a Pantheist, as Orpheus was, it is probable he would have idealized these things and created Greek lines. But believing in a distinct God, the supreme originator of all things, he was led to a worship of self-sacrifice and offerings, and needed no intellectual ideal, studiously celebrating and illustrating his own humanity; his service cultivated his moral rather than his æsthetical faculties. So, with a lavish hand, he appropriated the abundant beauty of nature, sometimes imitating its external expressions with his careful chisel, but generally conventionalizing them to suit the structural conditions of his capitals, spandrels, and panels, his crockets, belts, corbels, and the bosses of his soaring groins. The life of Gothic lines was in their sensuous liberty, though they were sometimes drawn with marvelous delicacy and refinement of feeling; the life of Greek lines was in their intellectual reserve. Those arose out of a religion of emotional ardor; these, out of a religion of philosophical reflection. Hence, while the former were free and picturesque, the latter were serious, chaste, and very human.

Therefore the line typical of the Gothic or lay spirit of the Middle Ages, like that of Roman, Byzantine, and Romanesque art, is still the full and sensuous line of life, as contrasted with the restrained and delicate line of Greek art. And yet when the noble artistic spirit of the thirteenth century interpreted in form the mysticism, the asceticism, the spiritual aspiration of the time, it sometimes seemed to confer upon this symbol a certain purity of movement which we do not find in the others. But it was not a Greek line; for the mediæval craft-spirit, however inspired, could never enter into that region of calm study and deep reflection from which alone could proceed the shafts of the Parthenon and the capitals of Athene Polias. And so the Greek lines slept in patient marble through the long Dark Ages, and no one came to awaken them into beautiful life again. No one, consecrated Prince by the chrism of nature, wandered into the old land to kiss the sleeping beauty into life, and break the deep spell which was around her kingdom.

Architecture promptly and exactly fulfilled its great function as a recorder of history, when, in the fifteenth and sixteenth centuries, it expressed in classic terms the revival of the literature and learning of Rome, the joyous emancipation of the mind from the gloom and doubt of the Middle Ages, and its new birth into a larger and fuller life. But when Brunelleschi and Alberti made the first attempts to utter this sentiment in Italian palaces and churches, archæology had not measured and restored the ruins of Rome, and æsthetics had not estimated their true value in the history of art. The Pantheon, the Colosseum, a few shattered remains of columns and entablatures, neglected, insulted, misunderstood, alone remained to testify to the magnificence of Rome. The first expressions of the Renaissance, at Florence, in the dome of the Cathedral, in San Lorenzo, San Spirito, and the Pitti palace; at Mantua, in San Andrea; at Rimini, in San Francisco, all following the impulse of classic Rome, as then imperfectly comprehended, committed the new movement irrevocably to architectural manifestations based on that line of life which had made the order of Jupiter Stator different from that of the Parthenon, and the portico of the Temple of Concord different from that of the Erechtheum. So wanton were the wreaths it curled in the sight of the great masters of that period, that they all yielded to its subtle fascinations and sinned like their mother Eve. Had such men as Bramante, Michel Angelo, San Gallo, Palladio, Scamozzi, Vignola, San Michele, Bernini, been inspired by the thoughtful lines of Greece, so catholic to all human moods, and so wisely adapted to the true spirit of reform, it is not too much to say that all subsequent art would have felt the noble impulse; that structure and the decoration of structure would have been once more reconciled, as they were in the temples of Greece and in the cathedrals of the Middle Ages. Pedantic formalism and academic rules would not have converted architecture into a series of elegant but conventional quotations from the classics, thus removing it from the sympathy of the people; and the refining influence of these great masters would have worked to far more profitable and various results. If the knowledge and ingenuity of modern times had been so directed, modern architecture would have been kept in a path of consistent development, adjusted to all the varying conditions of civilization.

The Gothic architecture of the early part of the fifteenth century was ripe for the spirit of healthy reform. It had been actively accumulating, during the progress of the age of Christianity, a boundless wealth of forms, a vast amount of constructive resources, and material fit for innumerable architectural expressions of human power. But in the last two centuries of this era the love which, here and there in the earlier period, as in the west fronts of Rheims and Chartres, had almost succeeded in giving a character of intellectual beauty to this sensitive art was gradually depraved and vulgarized by the pride of the workman; and a luscious and abandoned luxury of line led it farther and farther astray from the true path, till at last it became like an unweeded garden run to seed, and there was no health in it. In the year 1555, at Beauvais, the masonic workmen uttered their last cry of defiance against the old things made new in Italy. Jean Wast and François Maréchal of that town, two cathedral builders, said that they " had heard of the Church of St. Peter at Rome, and would maintain that their Gothic could be built as high and on as grand a scale as the antique orders of this Michel Angelo." And with this spirit they built a wonderful pyramid over the cross of their cathedral. But it fell in the fifth year of its arrogant pride, and this is the last we hear of Gothic architecture in those times. Over the picturesque ruins the spirits of the old conquerors of Gaul once more strode with measured tread, and began to set up their prevailing standards in the very strongholds of Gothic supremacy. These conquerors trampled down the true as well as the false in the mediæval *régime,* and the broad daylight of the Renaissance utterly extinguished that sole lamp of knowledge, which had given light to the ages of darkness and had kindled into life and beauty the cathedrals of Europe.

An architecture of mystic symbolism, of severe asceticism, of spiritual aspiration, of superstitious adoration and fear, could not live in this daylight of awakened intelligence. Whatever may have been its capacities for expansion, it could not have been continued in the new era without losing its character of truth as an exposition of history. The apostles of the Renaissance followed an impulse far more powerful than their own individual fancies when they rooted it out bodily, and planted instead an exotic of pagan Rome. It was the rebirth of an art *system,* which in its former existence had developed in an atmosphere of conquest. It taught them to kill, burn, and destroy all that opposed the progress of its triumph. It was eminently revolutionary in its character; and its reign, to all those multitudinous expressions of life and thought which had arisen under the Gothic era, was one of terror. Truly, it was a fierce and desolating instrument of reform.

It is a tempting theme of speculation to follow in the imagination the probable progress of a Greek instead of a Roman renaissance, into such active, intelligent, and imaginative schools as those of Rouen and Tours in the latter part of the fifteenth century, — of Rouen, with its Roger Arge, its brothers Leroux, who built there the picturesque Hôtel Bourgtheroulde, its Pierre de Saulbeaux, and all that legion of architects and builders who were employed by the Cardinal Amboise on his castle of Gaillon, — of Tours, with its Pierre Valence, its François Marchant, its Viart and Colin Byart, out of whose rich and playful craft-spirit arose the quaint fancies expressed in the older parts of the châteaux of Amboise and Blois, in which the old Gothic spirit seemed to have been refined and rejuvenated for a new and brighter career, which it was destined never to see. Such a renaissance would not have come among these venial sins of *naïveté,* this sportive affluence of invention, to overturn ruthlessly and annihilate. Its mission would inevitably have been, not to destroy, but to fulfill, — to develop from these strange results of human frailty and human power some new and as yet unimagined expression of prolific beauty, reflecting the progress of the race, and keeping the chain of architectural evidence unbroken by revolutions or revivals. For it seems logical that the principles of art, which nineteen centuries before had done a good work with the simple columns and architraves on the banks of the Ilissus, would under the guidance of love have made the arches and vaults and buttresses and pinnacles of that romantic civilization illustrious with even more eloquent expressions of refinement. For Greek lines do not stand apart from the sympathies of men by any spirit of ceremonious and exclusive rigor, as is undeniably the case with those which were adopted from Rome. They are not a *system* but a *sentiment,* which wisely directed might enter into the heart of any condition of society, and leaven all its architecture with a purifying and pervading power without destroying its independence, while an inflexible system could assume a position only by tyrannous oppression.

Yet when we examine the works of the Renaissance, after the system had become more manageable and acclimated under later Italian and French hands, we cannot but admire the skill with which the lightest fancies and the most various expressions of human contrivance were reconciled to the formal rules and proportions of the Roman orders. The Renaissance palaces and civil buildings of France, especially before the scholasticism and pedantry of the seventeenth century had quite overwhelmed the native spirit of romance, are so full of ingenuity, and the irrepressible inventive power of the artist moves with so much freedom and grace among the stubborn lines of that revived architecture, that we cannot but regard the results as a proof that the high creative power still lies dormant in the race, awaiting only for the opportunity or the initial force to express itself in a manner worthy of our civilization. We cannot but ask ourselves, If the spirit of those architects could obtain so much liberty under the restrictions of such an unnatural and unnecessary despotism, what would have been the result if they had been put in possession of the very principles of Hellenic art, instead of the dangerous and complex models of Rome, which were so far removed from the purity and simplicity of their origin? Up to a late day the great aim of the Renaissance has been to interpret an advanced civilization with the sensuous line; and, *so far as this line is capable of such expression*, the result has been satisfactory.

Thus four more centuries were added to the fruitless slumbers of Ideal Beauty among the temples of Greece. Meanwhile, in turn, the Byzantine, the Northman, the Frank, the Turk, and finally the bombarding Venetian, left their rude invading footprints among her most cherished haunts, and defiled her very sanctuary with the brutal touch of barbarous conquest. But the kiss which was to dissolve this enchantment was one of love; and not love, but cold indifference, or even scorn, was in the hearts of the rude warriors. So she slept on undisturbed in spirit, though broken and shattered in the external type, and it was reserved for a distant future to be made beautiful by her disenchantment and awakening.

In 1672 a pupil of the artist Lebrun, Jacques Carrey, accompanied the Marquis Ollier de Nointee, ambassador of Louis XIV., to Constantinople. On his way he spent two months at Athens, making drawings of the Parthenon, then in an excellent state of preservation. These drawings, more useful in an archæological than an artistic point of view, are now preserved in the Bibliothèque Nationale of Paris. In 1676, two distinguished travelers, one a Frenchman, Dr. Spon, the other an Englishman, Sir George Wheler, tarried at Athens, and gave valuable testimony, in terms of boundless admiration, to the beauty and splendor of the temples of the Acropolis and its neighborhood, then quite unknown to the world. Other travelers followed these pioneers in the traces of that old civilization. But in 1687, Königsmark and his Venetian forces threw their hideous bombshells among the exquisite temples of the Acropolis, and, igniting thereby the powder magazine with which the Turks had desecrated the Parthenon, tore into ruins that loveliest of the lovely creations of Hellas. It was not until the publishing of the famous work of Stuart and Revett on "The Antiquities of Athens," in 1762, that the world was made familiar with the external expressions of Greek architecture. This publication at once created a curious revolution in the practice of architecture, — a revolution extending in its effects throughout Europe. A fever arose to reproduce Greek temples; and to such an extent was this unintelligent revival carried out that at one time it bid fair to supplant the older renaissance. The spirit of the new renaissance, however, was one of mere imitation, and had not the elements of life and power to insure its ultimate success. It was a revival of formulas and not of principles. Nobody thought of applying the refining influences of the Greek lines to any structural forms save those known and used by the Greeks in their temples and stoas. The Roman arch, which had become indispensable, was not purified by the Greek spirit of truth, and the modern façade, without columns or pilasters, was not affected by it. No attempt was made to acclimate the exotic to suit the new conditions it was thus suddenly called upon to fulfill; for the *sentiment* which actuated it, and the love by which it was created, were not understood. It was the mere setting up of old forms in new places; and the Grecian porticoes and pediments and columns, which were multiplied everywhere from the models supplied by Stuart and Revett, and found their way profusely even into this new world, still stare upon us gravely with strange alien looks. The impetuous current of modern life beats impatiently against that cumbrous solidity of peristyle which sheltered well in its day the serene philosophers of the Agora, but which is the merest impediment in the way of modern traffic and modern necessities. The spirit of formalism, engendered by the old renaissance, took hold of the revived Greek lines, and stiffened them into acquiescence with a base mathematical system, which effectually deprived them of that life and reproductive power which belongs only to a state of artistic freedom. They were reduced to rule and deadened in the very process of their revival. The "Carpenters' Guides" in the opening of the nineteenth century explained to the builders the details of this system, and everywhere throughout our country the public buildings and private mansions of that period reproduced the forms of the Greek orders with unimaginative reiteration. The retired merchants of our Atlantic coast and the wealthy planters of Louisiana and Mississippi were housed behind stately porticoes of wood or stucco carefully imitated from the marble antiquities of Attica.

The Greek ideal, though strangely transplanted thus into the noise of modern streets, was not awakened from its long repose by the clatter and roaring of our new civilization.

As regarded the uses of life it still slept in petrifactions of Pentelic marble. And when those petrifactions were repeated in modern quarries, it was merely the shell they gave; the spirit within had not yet broken through.

Greek lines, therefore, owed their earliest revival to the vagaries of a capricious taste, and the desire to give zest to the architecture of the day by their novelty; it was not for the sake of the new life there was in them, nor of that refining spirit, by the application of which alone to modern design a wise re-birth of ancient love in art could be procured, without a sacrifice of everything else which has been inherited from the past. These lines are not an intolerant power, but a creative principle, by which all that is valuable of this inheritance may be preserved and adjusted to the complicated conditions of modern life. It is not surprising that some of the more modern masters of the Roman Renaissance, with whom that system had become venerable, from its universal use as the vehicle by which the greatest artists of the sixteenth and seventeenth centuries had expressed their thoughts and inspirations, regarded with peculiar distrust these outlandish innovations on the exclusive walks of their own architecture. For they saw only a few external forms which the beautiful principles of Hellenic art had developed to fit an old civilization; the applicability of these primary principles to the refinement of the architectural expressions of a modern state of society they could not of course comprehend. About the year 1786 we find Sir William Chambers, the leading architect of his day in England, in his famous treatise on "The Decorative Part of Civil Architecture," giving elaborate and emphatic expression to his contempt of that Greek art which had presented itself to him in a guise well suited to cause misapprehension and error. "It must candidly be confessed," he says, "that the Grecians have been far excelled by other nations, not only in the magnitude and grandeur of their structures, but likewise in point of fancy, ingenuity, variety, and elegant selection."

Two distinguished German artists — the one, Schinkel of Berlin, born in 1781, — the other, Klenze of Munich, born in 1784 — were children when Chambers uttered these treasonable sentiments concerning Greek art. Later, at separate times, these artists visited Greece, and so filled themselves with the feeling and sentiment of the art there, so consecrated their souls by the appreciative study of its divine love, that the patient Ideal at last awoke from its long slumbers, entered into the breathing human temples thus prepared, for it by the pure rites of Aphrodite, *and once more lived.* Thus in the opening years of the nineteenth century was a new and reasonable renaissance, not of an antique type, but of a spirit which had the gift of immortal youth, and uttered oracles of prophecy to these chosen Pythians of art.

Through Schinkel, the pure Hellenic style, only hinted at previously in the attempts of less inspired Germans, such as Langhaus, who embodied his crude conceptions in the once celebrated Brandenburg Gate, was fairly and grandly revived in the Hauptwache Theatre and the beautiful Museum and the Bauschule and observatory of Berlin. He competed with Klenze in a series of designs for the new palace at Athens, rich with a truly royal array of courts, corridors, saloons, and colonnades. But the evil fate which ever hangs over the competitions of genius was baleful even here, and the barrack-like edifice of Gütner was preferred. His latest conception was a design of a summer palace at Orianda, in the Crimea, for the empress of Russia, where the purity of the old Greek lines was developed into the poetry of terraces and hanging-gardens and towers, far-looking over the Black Sea. Schinkel was called the Luther of architecture; and the spiritual serenity which he breathed into the pomp and ceremonious luxury of the art of his day seems to give him some title to this distinction. Yet, with all the freedom and originality with which he wrought out the new advent, he was perhaps rather too timid than too bold in his reforms, — adhering too strictly to the original letter of Greek examples, especially with regard to the orders. He could not entirely shake off the old incubus of Rome.

And so, though in a less degree, with Klenze. When, in 1825, Louis of Bavaria came to the throne, he was appointed government architect, and in this capacity gave shape to the noble dreams of that monarch, in the famous Glyptothek, the Pinakothek, the palace, and those civil and ecclesiastical buildings which render Munich one of the most monumental cities of Europe. It was his confessed aim to take up the work of the Renaissance artists, having regard to our increased knowledge of that antique civilization of which the masters of the sixteenth century could study only the most complex developments, and those models of Rome which were farthest removed from the pure fountain-head of Greece. "To-day," he said, "put in possession of the very principles of Hellenic art, we can apply them to all our actual needs, — learning from the Greeks themselves to preserve our independence, and at the same time to be duly novel and unrestrained according to circumstances." These are certainly noble sentiments; and one cannot but wish that when, in 1830, Klenze was called upon to prepare plans for the grand Walhalla of Bavaria, he had remembered his sublime theory and worked up to its spirit, instead of recalling the Parthenon in his exterior and the Olympian temple of Agrigentum in his interior. The last effort of this distinguished artist was the building of three superb palaces for the museum of the Emperor at St. Petersburg, finished in 1851.

The seed thus planted fell upon good ground and brought forth a hundred-fold. This revival of Hellenic principles is still infusing life into modern German designs; and so well are these principles understood that architects do not content themselves with the mere reproduction of that narrow range of motives which was uttered in the temples of heroic Greece. By the very necessity of the Greek line, they are rendered catholic and unexcluding in their choice of forms, but fastidious and hesitating in expressing them in this new language of art. This wholesome restraint is recognizable in much of their recent work, and it undoubtedly is a result of their Greek studies. But inasmuch as these are pursued with the true German spirit of scientific precision and system, it cannot be denied that modern German art has still a character of pedantry. The national genius aims at correctness in art and archæology, as well as in science and philosophy. It is not discursive or imaginative, but essentially scholastic; it respects the authority of formulas and the discipline of dogmas. It is not original but exact. Within these limitations the best work of the nation is admirable, but not often interesting. It nearly always bears the stamp of conscientious study; but it is rarely brilliant. For its best qualities it is clearly indebted to those two illustrious pilgrims who brought back from the land of epics, not only the scallop-shells upon their shoulders, but in their hearts the consecration of ideal beauty.

According to the usual custom, in the year 1827 a scholar of the Ecole des Beaux Arts in Paris, having achieved the distinguished honor of being named *grand pensionnaire* of architecture for that year, was sent to the Académie Française in the Villa Medici at Rome, to pursue his studies there for five years at the expense of the government. This scholar was Henri Labrouste. While in Italy his attention was directed to the Greek temples of Pæstum. Trained, as he had been, in the strictest academic architecture of the Renaissance, he was struck by many points of difference between these temples and the Palladian formulas which had hitherto held despotic sway over his studies. In grand and minor proportions, in the disposition of triglyphs in the frieze, in mouldings and general sentiment, he perceived a remarkable freedom from the restraints of his school,— a freedom which, so far from detracting from the grandeur of the architecture, gave to it a degree of life and refinement which his appreciative eye now sought for in vain among the approved models of the Academy. Studying these new revelations with love and veneration, it was not long before the pure Hellenic spirit, confined in the severe peristyles and cellas of the Pæstum temples, entered into his heart with all its elastic capacities, all its secret and mysterious sympathies for the new life which had sprung up during its long imprisonment in those stained and shattered marbles. Labrouste, on his return to Paris in 1830, surprised the grave professors of the Academy, Le Bas, Baltard, and the rest, by presenting to them, as the result of his studies, carefully elaborated drawings of the temples at Pæstum. Witnessing, with pious horror, the grave departures from their rules contained in the drawings of their former favorite, they charged him with error, even as a copyist. True to their prejudices, their eyes did not penetrate beyond the outward type, and they at once began to find technical objections. They told him, never did such an absurdity occur in classic architecture as a triglyph on a corner! Palladio and the Italian masters never committed such an obvious crime against propriety, nor could an instance of it be found in all Roman antiquities. It was in vain that poor Labrouste upheld the accuracy of his work, and reminded the Academy that among the Roman models no instance had been found of a Doric corner, — that this order occurred only so ruined that no corner was left for examination, or in the grand circumferences of the Colosseum and the Theatre of Marcellus, where, from the nature of the case, no corner could be. The professors still maintained the integrity of their long-established ordinances, and, to disprove the assertions of the young pretender, even sent a commission to examine the temples in question. The result was a confirmation of the fact, the ridicule of Paris, the consequent branding of the young artist as an architectural heretic, and a continued persecution of him by the Ecole des Beaux Arts. Undaunted, however, Labrouste established an *atelier* in Paris, to which flocked many intelligent students, sympathizing with the courage which could be so strong in the conviction of truth as to brave in its defense the displeasure of the powerful hierarchy of the school.

Thus was founded a new renaissance in France at a time when the genius and ingenuity of the nation was beginning to chafe under the severe restrictions of the old academic formulas; and in this genial atmosphere Greek lines began to exercise an influence far more thorough and healthy than had hitherto been experienced in the whole history of modern art. To the lithe and elegant fancy of the French this revelation was especially grateful. It seemed in a measure to restore to them the freedom of which they made such magnificent use in the time of Francis I., before the Renaissance had been academically formalized by the Mansarts and their successors. For the youth of this nation soon learned that in these newly opened paths their invention and sentiment, so long straitened and confined within the severe limits of the old system, could move without irksome restraint, and at the same time be preserved from licentious excess by the delicate spirit of the new lines. Thus natural fervor, grace, and fecundity of thought found here a most welcome outlet.

For some time the designs of the new school were not recognized in the competitions of the Ecole des Beaux Arts; but when, in the course of nature, some two or three of the more strenuous and bigoted professors of Palladio's golden rules were removed from the scene of contest, the new cult was received at length into the bosom of the architectural church, and now it may be justly deemed a distinctive architectural expression of French art.

Labrouste was not alone in his efforts; but Duban and Constant Dufeux seconded him with genius and energy. Most of the important buildings which have been erected in France within the last fifty years have either been unreservedly and frankly affected in their general architectural character by the Greek lines, or have been refined by more limited applications of Hellenic principles. Even the revived mediæval school, which, under the distinguished leadership of M. Viollet le Duc and the lamented M. J. B. A. Lassus, was strengthened to a remarkable degree in France, and which shared with the Greek idea the displeasure of the Academy, — even this tacitly acknowledged the power of Greek lines, and instinctively suffered them to purify, to a certain degree, the old grotesque Gothic license. One of the most remarkable peculiarities of this school was, that it seemed to encourage the expression of individual traits in the architects; to give fluency and ease to architecture, and to free the spirit of the designers from the impediments of conventionalism. Indeed, when Greek lines were first revived in the Paris *ateliers*, the architects were so much impressed by the freedom which the use of these lines gave to all the processes of design, when compared with the restrictions of practice under the Roman academic system, that the new dispensation of art was called a style, and christened Romantique, to distinguish it from what was conventionally called classic. Of course this freedom, in insincere hands, was frequently abused, and this abuse made it all the more evident that the only acceptable service of Greek lines is one of love, and not of affectation or imitation. The ordinary decorations of windows and doors were not confined to conventional shapes, as of yore, but were idiosyncratic, according to the degree in which they were influenced by the Greek spirit. If the designer had a thought to express, his Greek lines enabled him to put it in architectural form, just as a poet attunes *his* thought to the harmony and rhythm of verse. Antique prejudices, bent into rigid conformity with antique rubrics, were often shocked at the strange innovations of these new dissenters from the faith of Palladio and Philibert Delorme, — shocked at the naked humanity in the new works, and would cover it with the conventional fig-leaves prescribed in the homilies of Vignola. Laymen, accustomed to the cold architectural proprieties of the old renaissance, and habituated to the formalities of the five orders, the prudish decorum of Italian window-dressings and pediments and pilasters and scrolls, were surprised to see ideas at length so clearly set forth in architectural forms that the intention of the building in which they occurred was at once patent to the most casual observer, and the story of its destination told with the eloquence of a poetical and monumental language. All revolutions have proved how hard it is to break through the crust of custom, and this was no exception to the rule; yet in justice it must be said that every intelligent mind, every eye possessing the "gifted simplicity of vision," to use a happy phrase of Hawthorne's, recognizes the emancipation and enlargement effected by the application of the true Greek spirit to the arts of design, and sees in it a clarifying and renovating influence, clearing away the dust and cobwebs which ages of prejudice have spread thickly around the magnificent art of architecture.

Unlike the unwieldy and ponderous classic or Italian systems, whose pride cannot stoop to anything beneath the haughtiest uses of life without being broken into the whims of the grotesque and *rococo*, the Greek renaissance has already proved that it may be applied with graceful ease to the most playful as well as the most serious employments of art. It has decorated the perfumer's shop on the boulevards with the most delicate fancies woven out of the odor of flowers and the finest fabrics of nature, and, in the hands of Labrouste, has built the great Bibliothèque Ste. Geneviève, the most important work with pure Greek lines, and perhaps the most exquisite, while it is one of the most serious, of modern buildings. The learning exhibited in this composition does not make it pedantic, its careful simplicity of *motif* does not weaken its interest, nor does its refinement and purity destroy its power. The neo-Grec has also been used with especial success in funereal monuments. Structures of this character, demanding earnestly in their composition the expression of human sentiment, have hitherto been in most cases unsatisfactory, as they have been built out of a narrow range of renaissance, Egyptian, and Gothic motives, originally invented for far different purposes, and since then classified, as it were, for use, and reduced to that inflexible system out of which have come the formal restrictions of modern architecture. Hence these formulas have never come near enough to human life, in its individual characteristics, to be plastic for the expression of those emotions to which we desire to give the immortality of stone in memory of departed friends. The neo-Grec, however, confined to no rigid types of external form, out of its noble freedom is capable of giving "a local habitation and a name" to a thousand affections which hitherto have wandered unseen from heart to heart, or been palpable only in words and gestures which disturb our sympathies for a while and then die. Probably one of the most remarkable indications of this capacity as yet shown is contained in a tomb erected by Constant Dufeux in the Cimetière du Sud, near Paris, for the late Admiral Dumont d'Urville. This structure includes in its outlines a symbolic expression of human life, death, and immortality, and in its details an architectural version of the character and public services of the distinguished deceased. The finest and most eloquent resources of color and the chisel are brought to bear on the work; and the whole, combined by a very sensitive and delicate feeling for proportion, thus embodies one of the most expressive elegies ever written. The tomb of Madam Delaroche, *née* Vernet, in the Cimetière Montmartre, by Duban, is another remarkable instance of this elastic capacity of Greek lines;

and though taken frankly, in its general form, from a common Gothic type, its chaste and graceful freedom from Gothic conventions of detail gives it a life and poetic expressiveness which must be exceedingly grateful to the love which commanded its erection.

Paris thus affords us, in some of its modern architecture, a happy proof of the inevitable reforming and refining tendencies of the abstract lines of Greece, when properly understood and fairly applied. Under their influence old things have been made new, and the coldness and hardness of academic art have been warmed and softened into life. Through the agency of the Greek school, perhaps more new and directly symbolic architectural expressions have been uttered within the last forty years than previously since the beginning of the seventeenth century. Like the gestures of pantomime, which constitute an instinctive and universal language, these abstract lines, coming out of our humanity and rendered elegant by the idealization of study, are, it is hoped, restoring to architecture its highest capacity of conveying thought in a monumental manner.

One of the most dangerous results of that eclecticism which the advanced state of our archæological knowledge has made the principal characteristic of modern design consists in the fatal facility thus afforded us of availing ourselves of vast historical resources of forms and combinations ready made to suit almost all the exigencies of composition, as we have understood it. The public has thus been made so familiar with the set variations of classic orders and Palladian windows and cornices, with Venetian balconies and arcades, with all sorts of Spanish, Italian, English, French, and Dutch interpretations of the classic theme, with all manner of Gothic chamfers and cuspidations and foliations, and the other conventional symbols of architecture, which undeniably have more of *knowledge* than *love* in them, — so accustomed have the people become to these things, that the great art of which these have been the only language now almost invariably fails to strike any responsive chord in the human heart, or to do any of that work which it is the peculiar province of the fine arts to accomplish. Instead of leading the age, it seems to lag behind it, and to content itself with reflecting into our eyes the splendor of the sun which has set, instead of facing the east and foretelling the glory which is coming.

GREEK LINES.

Architecture, properly conceived, should always contain within itself a direct appeal to the sense of fitness and propriety, the common-sense of mankind, which is ever ready to recognize reason, whether conveyed by the natural motions of the mute, or the no less natural motions of lines. Now history has proved to us, as has been shown, how, when the eloquence of these simple, instinctive lines has been used as the primary element of design, great eras of art have arisen, full of the sympathies of humanity, immortal records of their age. It cannot be denied, on the other hand, that our eclectic architecture, popularly speaking, is not comprehended, even by the most intelligent of cultivated people; and that, so long as it is contented to be merely reminiscent and archæological, the inevitable rapid development of structural science and new materials, and the changes in our social conditions, which necessarily must impose new forms upon buildings, cannot be properly and adequately recognized in art. When architects, instructed by the past, and guided by principles, and not by the prejudices of schools, seek to give these new forms a consistent decorative character, they are actually creating an architecture belonging to our times and to our people.

The French of to-day are apparently not so loyal to the Greek revival as they were a few years ago; possibly they have been discouraged by the results of one or more false starts. Yet I think it may be safely said that the characteristics of the best contemporary art of Paris are controlled by a Greek spirit, though it still has to struggle against academical influences, long established, and inhospitable to any innovation which does not recognize the venerable formulas of Vignola.

Let our artists turn to Greece, and learn how, in the meditative repose of that antiquity, these ideals arose to life beneficent with the baptism of grace, and became visible in the loveliness of a hundred temples. Let them there learn how in our own humanity is the essence of form as a language, and that *to create*, as true artists, we must know ourselves and our own distinctive capacities for the utterance of monumental history. After this sublime knowledge comes the necessity of the knowledge of precedent. The great Past supplies us with the raw material, with orders, colonnades and arcades, pediments, consoles, cornices, friezes and architraves, buttresses, battlements, vaults, pinnacles, arches, lintels, rustications, balustrades, piers, pilasters, trefoils, and all the innumerable conventionalities of architecture. It is plainly our duty not to revive and combine these in those cold and weary changes which constitute much of modern design, but to make them live and speak intelligibly to the people through the eloquent modifications of our own instinctive lines of life and beauty.

The riddle of the modern Sphinx is, How to create a new architecture? And we find the Œdipus who shall solve it concealed in our own hearts.

THE GROWTH OF CONSCIENCE IN MODERN DECORATIVE ART.

SELF-EXAMINATION has become one of the characteristic instincts of modern civilization. It was not long ago that Carlyle described this instinct as a sort of moral dyspepsia, prevailing more or less absolutely in all the grades of society. However this may be, it is true that, unlike our forefathers, we take nothing for granted. The religious beliefs, the social traits, the manners and customs, which we have inherited from them are subjected to analysis and discussion. A new electric searchlight reveals the errors of conservatism. The result of this study is the establishment of certain reasonable but not unchanging types, with which, in the conduct of life, according to our several lights, we seek to establish a conscientious conformity. Concerning art, however, for various reasons which we shall presently consider, there has been until lately a reluctance to bring to bear upon it any such reorganizing and revolutionary tendencies.

Hitherto, when those of us who have been engaged in works of architectural or decorative design have undertaken, in the modern spirit, to analyze our motives in any succession of cases, we have found that the standard of excellence by which we would measure our work, the ideal which we would approach, has been, so far as the form at least was concerned, inconstant and for the most part capricious. These variations of style have not occurred according to any known law. Our art seems to have been in great degree controlled by some power outside of ourselves, — by the prevalence of a fashion, or by the influence of some successful master. We have found it convenient and comfortable to accept the dictates of this power without questioning, and our standard has been set up successively in ancient Greece or Rome, in mediæval France, England, or Italy. At one time it has held to some phase of the Renaissance; at another, Romanesque art has presented itself as the only recuperative power; at still another it has been absolute as to its Gothic: "all these by turns and nothing long." Its caprice has been curious and unaccountable, and not at all in accordance with the modern spirit in other walks of intelligence.

This vacillation of the type which has prevented modern architecture from developing a style, in the accepted sense of the word, is the natural result of the increase of our knowledge of form, as developed in historical architecture, and the growth of the archæological spirit. Unlike any of our predecessors in art, we have been seriously embarrassed by the unbounded range and variety of precedent at our command. There is no phase of historical art which we have not studied; wheresoever and howsoever humanity has expressed itself in forms of art, these forms are at our fingers' ends, and are ready to seduce us this way or that, according to our mood. The mind of the designer is preoccupied by innumerable favorite *motifs* derived from every side and every era, each associated with some phase of ancient life, and sanctified or sweetened by ancient traditions; each with a value aside from intrinsic picturesqueness, beauty, or quaintness, and all contending for new expression. Whether he has been engaged upon a composition of architecture or upon a composition of decoration,— which also is architecture, or the completion and fulfillment of it, — his energy has been concerned first, perhaps, with the choice of types agreeably to the caprice or fashion of the moment; next, with the degree of precision with which he is to follow them when chosen; and, finally,— with such reserve of force as might be at his disposal after these exhausting processes,— with the adjustment of his chosen forms to his needs according to his best ingenuity and skill. Under these circumstances, the modern process of design, whether this exact order of proceeding has been followed or not, must be a complicated one, and must differ fundamentally from all which have preceded it. The exact character of this difference it is important for us to understand at the outset, to the end that we may the better comprehend the new and strange conditions under which art is developed in these modern days.

The Greek architect of the time of Pericles had before him a fixed and sacred standard of form. There were probably dim traditions from his Pelasgic ancestors, and from Syria and Egypt. These were the only styles or forms that he knew, and his own had been developed from them into a hieratic system. He had no choice; his strength was not wasted among various ideals; that which he had inherited was a religion to him. The simple cella with a portico or peristyle,—this was all; he had no wants or ambitions beyond this; it satisfied all his conditions of art. But he shared in the intense intellectual activity of his fellow-citizens; his art had been developed in the same atmosphere as the philosophy of Plato and Aristotle, the drama of Æschylus, Sophocles, and Aristophanes. He was content with nothing but absolute perfection. Undiverted by side issues as to the general form of his monument, undisturbed by any of the complicated conditions of modern life, he was able to concentrate his clear intellect upon the perfection of his details; his sensitiveness to harmony of proportion was refined to the last limits; his feeling for purity of line reached the point of a religion. Design, in fact, was studied seriously in the midst of a deep silence, like an act of worship. In the previous essays an attempt has been made to indicate how this spirit of refinement and delicacy operated in the development of style, and how, upon its rediscovery and application to our modern works, depends in a great degree the rehabilitation of our architecture.

In like manner, many centuries subsequent the monkish builders developed the Christian temple in the cloisters of Cluny. All that they knew of style had been developed in a direct line of descent from Gallo-Roman traditions, and they, like the Greek, were undisturbed by any knowledge of conflicting forms. Their art was thus kept in the track of consistent progress, and developed with purity and irresistible force.

So it was with all the intermediate builders. So it was when the Taj Mahal was built in Agra. So it was wherever there grew a pure style. So it was even after the period of the Renaissance. The development of styles continued strong and steady until archæology began to revive, classify, and make known to the world, as a contribution to history, the various methods and forms which were pursued and invented by old civilizations in the erection of their temples, tombs, and palaces. Then there followed a confusion of tongues which has lasted until our day.

From all this it necessarily follows that the distinctive characteristic of our modern art is the absence of a fixed standard of forms. It is eclectic, and apparently has not encouraged us to reach convictions as to forms or styles. At all events, there are few architects or designers, in this country at least, who are content to confine themselves to the exclusive development of any one particular set of forms, as Gothic, or Romanesque, or Renaissance, and voluntarily to shut themselves off from the rest of their inheritance of beautiful things; and wherever any such exist their neighbors are not so confined. In this particular we do not work together with any characteristic unity of sentiment. All the decorative arts are subjected to the same dissipation of forces. At the same moment we are designing and painting Greek vases; decorating Japanese screens; constructing furniture according to our reminiscences of the Gothic of the Edwards, or of the Renaissance of the Jameses, of Queen Anne, or of the Georges; covering our walls with designs suggested by the stuffs of Florence and of the exhaustible East, by the brocades of France, by the stamped leather of Venice, with arabesques and conceits from all the styles; and with these we decorate the interiors of houses which on the outside have been inspired originally from traditions of every era of art, as set forth in books, prints, and photographs innumerable.

It is therefore a common reproach against the arts of to-day that they are discursive, without convictions or enthusiasm; that our depth is shallowed in many channels; that we produce many and not great things; that in painting we have no masterpieces like those of Italy in the fifteenth century, or of Flanders in the sixteenth; that in sculpture the ideal of the Greek marbles, though shattered and defiled, is to us absolutely unapproachable, not in execution only, but in comprehension; that in architecture we cannot reproduce the perfection, the purity, and perfect fitness of the Greek forms, the grandeur and extent of those of the Roman Empire, the idealism, the enthusiasm, the consistent and powerful development, of the religious works of the thirteenth century, the elegance and refinement and self-control of the Italian masters of the fifteenth century, or the innocent exuberance of the French builders in the sixteenth; that in the fictile arts we cannot approach the French and Italian potters of that era; that in fabrics we are still far excelled by the Orientals, and by the products of mediæval looms; that in furniture, for fertility of design, for perfection of execution, for richness of carving, we are surpassed by the Philibert de l'Ormes, the Le Pautres, the Boules, of France, in the fifteenth and sixteenth centuries, by the Gibbons and the Chippendales of England in the eighteenth. In like manner we know that antique gems and intaglios, Etruscan jewels, boxes, fans, and bronzes of Japan, ironmongery of Nuremberg, — these, in their several departments of art, are the despair of modern workmen; that in no respect of art do we exceed our progenitors. It would seem, in fact, as if our knowledge, our ingenuity, our industry, had swamped our art, — as if our art were in a condition, if not of eclipse, certainly of hopeless anarchy, and this while its patrons were apparently never so rich, never so numerous, never so ready.

In the presence of these new and incongruous elements there has come into existence, in the latter part of the nineteenth century, a spirit hitherto unknown in the history of art, — the spirit of self-consciousness in the modern artist. The innocence and *naïveté* of the older day have gone by, never to return. Our ancestors perhaps "builded better than they knew." But we can never do a good thing by accident. Each of us, in whatever style he may work, must necessarily impress himself upon his design. We can never be quite lost in the style which we have chosen. A new, subjective, personal element has thus been born into art. This self-conscious spirit began to be felt when the necessity of making choice among several types or styles was first imposed upon the artist; this choice implying the idea of self-justification, and giving an added sense of personal responsibility, which has naturally grown with the increase of our knowledge. During the existence of a prevailing or exclusive style, as in any time previous to the middle of the last century, there was far less scope for individuality of expression than now, when the necessity of making choice among many styles and among innumerable *motifs* constantly compels the designer to a review of his own resources and acquisitions, and to a special adjustment of them to new conditions of structure, use, and material. The new labors thus imposed upon him of selecting or rejecting with discretion, of combining and modifying, necessarily result in an expression of his own peculiarities of thought and habits of mind, which would have been impossible to a Greek of the time of Pericles, or to a Frenchman of the time of St. Bernard.

Hermogenes and Callicrates, Apollodorus and Vitruvius, Viellard de Honcourt, Robert de Luzarches, and William of Wyckham, — each of these concerned himself with the development of a type of form, and carried it on one step further towards perfection. In this type their individuality was lost. They and their brethren are therefore but the shadows of names. Erostratus the fool, who burnt the Temple of Diana at Ephesus, is far better remembered in history than Ctesiphon, the architect who built it. Ctesiphon, though a great artist, was but the agent of a process of development in style; his work was rather a growth than a creation. But Sir Charles Barry, Alfred Waterhouse, Charles Garnier, Karl Friedrich Schinkel, of our time, built their monuments in the Houses of Parliament at London, the Law Courts at Manchester, the New Opera at Paris, and the Royal Theatre at Berlin; these buildings, and all other conspicuous monuments of modern times, are full of the personality of their authors, because they are rather creations than growths. Even so late as the sixteenth century the Renaissance palaces of Italy, built by Vignola, Scamozzi, Serlio, and Palladio, who knew no models but those of Rome, do not betray the personal characteristics of their designs to the same extent and in the same manner as do the neo-Greek works of Henri Labrouste in Paris, the modern Gothic of Scott, Burgess, and Street respectively, the modern Greek of the Scotch Thompson in Edinburgh, the "Queen Anne" revival of Norman Shaw, the Romanesque revival of our own Richardson, and so on through a host of more or less illustrious contemporaries, most of them changing their styles from time to time according to their moods.

Confused amongst a multiplicity of types, we impress upon our work a certain effect of archæological pedantry or breathless effort; we are either affected and coldly precise, or we overcrowd our designs with detail; our greatest and most difficult virtues, therefore, are sincerity, reserve of force, self-denial, simplicity, repose. The artists of antiquity found simplicity and repose in mere fidelity to a rigid standard, — a fidelity untempted by the discoveries of archæology, and easy because of the purity and perfection of the type. Their ideal was a divinity; their service to this divinity was worship and obedience. Our ideal is a museum of heterogeneous and beautiful forms, and our service to it is selection, rejection, adaptation, analysis, discussion, classification. Indeed, the modern artist is not the servant of his ideal; he properly seeks to be its master. Whenever, like a mediæval artist, he tries to render obedience to the ideal, the very perfection of his knowledge betrays him. However faithful he would be to his selected type of forms, he must needs breathe into it a spirit quite his own. If he would reproduce in his modern work the strong Gothic of the early Cistercian abbeys, he remembers also the refinements of Giotto in the Campanile of Florence. If he would imitate the elegant exuberance of the Roman Ionic, he cannot forget the fine chastisement of invention in the Ionic of the portico of Minerva Polias on the Acropolis of Athens. Thus his work is sophisticated by his knowledge. He is like an actor playing a part. He cannot conceal his effort. He is self-conscious.

In this way the modern spirit of self-examination, of which I have spoken, is gradually applied to art. The application of a rule of morality to the arts of design follows naturally, and is in exact harmony with the modern spirit of culture. I desire to treat of this growth of conscientiousness as the quality most characteristic of the art of to-day, — a quality which until now has never made its appearance in the decorative arts, and from which the most happy results may be reasonably anticipated; without which, in fact, these arts will become mere antiquarianism, destitute of soul or inspiration.

Conscientiousness, as applied to architecture, corrects and supplements feeling; it does not take the place of feeling. It regards this art, not as a business, or an amusement, or as a service of grace and poetry merely, but as a duty, carrying with it certain moral responsibilities like any other duty. This is a modern idea; it consists in the desire to establish some constant and conscious standard, by the observance of which, in the midst of the enormous and complicated demands made upon the decorative arts in our day, in the midst of the embarrassing accumulation of available and conflicting precedents, in the midst of the new materials, new inventions, new creeds, new manners and customs, constantly presenting themselves, a new art made up of many arts may be formed and kept from anarchy and confusion.

Now how is this newly awakened conscience of the architect or decorative designer applied to his work? In what way does it correct his taste or affect his artistic instincts?

It has been discovered that, in every great era of art, material has been used according to its natural capacities: by the consistent use of such natural capacities the arts have approached perfection; by their abuse they have inevitably declined. Thus, as regards architecture, in a district which produced granite alone, the prevailing style would submit to certain modifications to suit the conditions of the material: the mouldings would be few and large, the sculpture broad and simple, depending rather upon outline than upon detail for its effect. In places where the stone was easily worked, the mouldings and carvings would be more frequent. Where fine marbles were available, the architecture would be delicately detailed, and affect a quality of refinement impracticable under other conditions. Where colored marbles abounded, the wall surfaces would be veneered with them in patterns, and designs in tarsia would become frequent. Where clay only prevailed, there would arise an architecture distinctly of brick and terra cotta. If the stone of the district was coarse and friable, it would be used in rough walls, covered with a finish of cement or plaster, which in its turn would create a modification of style priding itself upon its smoothness of surface, its decoration by incisions and fine moulding and applied color. Thus, Egyptian art was, in some of its most characteristic expressions, an art of granite; the mediæval arts of France and England were mostly arts of limestones and sandstones of various qualities; the art of Greece was an art of fine marble; that of North Italy was an art of baked clay; that of Venice and Florence was distinguished for its inlays of semi-precious marble; that of Rome, as her monuments were a part of her political system, and were erected all over the Roman world as invariable types of her dominion, was an art of coarse masonry, in whatever material, or of concrete, covered with moulded plaster or with thin veneers of marble. In like manner, forms executed in lead were different from forms executed in forged iron. Forms cast in moulds were different from forms forged or wrought with the chisel. Forms suggested by the functions and capacity of wood were quite different from any other.

It is no less true, however, as is well known, that in their origins the Greek styles bore reminiscences of the primitive arts. The pylons of Egypt recalled the structures of mud and reeds which preceded them; the marble temples of Greece remembered the wooden frames of the primeval buildings;

and the early Gothic of France received its first decorations from hints in Oriental fabrics displayed by the Venetian merchants in the markets of Limoges. But when these styles reached perfection, the materials actually used in construction, and their capacity for legitimate expression, had been fully developed in each case: granite no longer resembled mud; marble no longer was fashioned into wooden forms; and limestones and sandstones were decorated, not like stuffs, but in such a manner that from a drawing of an ornament one could almost predicate the quality and grain of the material for which it was designed and in which it was executed.

This quality is called truth of material. There are also truth of construction and truth of color. They all are arrayed against imitations, against producing in one material forms invented for another, against concealment of devices of construction; in short, against sham work of any kind. Thus a certain master lays down this dogma: "A form which admits of no explanation, or which is a mere caprice, cannot be beautiful; and in architecture, certainly, every form which is not inspired by the structure ought therefore to be neglected."[1]

[1] Viollet-le-Duc.

Such doctrines as this have been so often preached in the literature of the times that they have become commonplace. They sufficiently indicate the conscientious tone of public sentiment as regards art, and the designers silently but diligently endeavor to meet the demand for a moral art with all the accepted devices of truthful work. About the middle of the nineteenth century, when the idea of morality in art was new, and its first disciples were flushed by the consciousness of their own piety, the dogma brought about results in practice which clearly indicated the supersensitive condition of the artistic mind at that time. We had an era of moral furniture, of which Mr. Eastlake was fortunate enough to be the prophet; but as the conclusions of the dogma were too rigid, its requirements too exacting, its illustrations of the principle of truth of material and truth of construction too literal and narrow, its productive power was speedily exhausted. There were doctrinaires, precisionists, *petits maîtres*, formalists, in this conscientious movement in the arts, as in every other new intellectual activity; they were ready to push the newly discovered principles to conclusions too absolute and mechanical, and to expose our arts to the danger of a recoil. Thus the "Eastlake furniture," which excluded curved lines on principle; which made the manner of construction — the joiners' part of it — more important than the designers'; which elevated the mortise and tenon to the dignity of a principle of art, by very reason of its great show of honesty, like any other ostentation of morality, soon palled upon the senses. With our inexhaustible inheritance of forms in which curved lines do appear, in which the idea of the designer is of more importance than the device of the cabinet-maker, we cannot remain long content with such pious exclusions.

But with all this, the conscientious spirit once aroused in art is not likely to be put to sleep again until a great work has been done. The designer would not quiet it if he could, for it gives to his work a new significance and power; it enables him to defend it by saying, "This composition of lines or of colors I am satisfied with, not merely because it gives me a sensuous gratification; not merely because it recalls this or that *motif* in some of the *chefs-d'œuvre* of art; nor because it reminds me of certain historic forms rendered precious by traditions and long use; not because it copies nature exactly; nor because it follows a certain accepted fashion of the day, — but because I know it is right. And why? I have reasoned about it, and can explain it by an appeal to your intellect. It belongs in its place, and accomplishes its object with a directness which could not be reached by mere intuition. It is not a mere matter of taste, concerning which there is no disputing. I cannot do otherwise than I have done and remain true to the conditions of my art. My forms are developed out of the necessities of my problem; they are not chosen because they are beautiful only, but because they are fit. Indeed, they would not be beautiful for my use if they were not fit. I have been taught by experience to distrust my own intuitive fancies and predilections for this or that form, for this or that style; they seduce me from the truth. I have been taught to discipline my resources; to subject them to critical analysis and discussion within my own mind before using them; to lop off what is irrelevant to my theme; to give greater emphasis here; to distract attention there; to harmonize the whole with the especial demands of my subject. I find that these conscientious processes, so far from weakening my fancy, so far from diminishing the interest of my work, in reality make my resources of design more available for my use, and render my compositions far more beautiful than any that I did before I had taught myself to reason. I have learned the artistic value of sincerity, simplicity, and self-denial. I not only feel, but I understand."

Before the latter half of the nineteenth century such language as this would have been impossible, but now it simply illustrates a common thought of the modern designer undertaking to create works of art; it illustrates a growing spirit in all the decorative arts. The masters of the great historical styles were constrained to concentrate their attention on the purification and development of a single type; we, among innumerable conflicting types, instead of vacillating, are beginning to reason, to extract from them principles of design, to apply rules of æsthetics. It is evident that no one can invent a new set of forms, conceived on new principles, which shall obliterate the memory of all that archæology has given us, and therefore that, whatever style may succeed in obtaining recognition as modern, it must have its roots somewhere in the past. With its vast inheritance, architecture must be more dependent on derivation than invention. It is equally evident that our resources of precedent will increase with the progress of time. Where, then, are we to look for a remedy for the increasing embarrassment of our knowledge? What can relieve us from an anarchy of forms on the one hand, or from the ignoble domination of a series of unreasonable, capricious fashions or revivals on the other?

It seems logical to infer that, as in the sciences the accumulation of knowledge never has been regarded as an affliction, so in art the accumulation of precedents from Greece, Rome, and Byzantium, from Egypt and Syria, from the Orientals and Spanish Moors, from mediæval Christendom, from the masters of the fifteenth century, from the châteaux and palaces of the Renaissance, from the revivals and rehabilitation of all those forms by our many-sided contemporaries, — this abounding wealth should hardly prove an embarrassment to us unless we are unfit to use so precious a heritage.

With this heritage we have tried all sorts of experiments. We have, for example, tried the effect of arbitrary exclusions. The time is not far distant when the world of art was divided into hostile camps, some holding to one set of precedents and regarding all others as misleading and pernicious, the rest considering that safety resided only in the very forms rejected by their competitors, — some for Gothic and some for classic. Fifty years ago, two architects could not meet without a quarrel. It was the "battle of the styles." We have tried this in architecture and in the other decorative arts, but have found that under such divisions we have made no progress. We have also, in turn, tried indifference as to the quality of the precedent, and masqueraded now in one dress and now in another, curious only in the perfection and accuracy of our copying; in other words, we have tried pure archæology, and found that it could not satisfy the cravings of the artist to create.

We are now at last beginning to learn that this great inheritance of forms is in fact the legitimate language of our art, copious, rich, suggestive, sufficient to all our moods; valuable to us, not for the sake of its own words and expressions and phrases, but because of its usefulness in enabling us the more fully and elegantly to express our own thoughts and the ideas which belong to our time.

To obtain success in the decorative arts, according to this new light, there must now be added to the qualification of the artist a new and hitherto unknown element, that of research and learning. The artists of classic and romantic times expressed themselves with ease in their own vernaculars, without knowledge of any other mediums of expression. We of modern times can hardly be said to have vernacular tongues, but must say what we have to say in a composite of many and various elements. We cannot use this composite without a process of reasoning. In expressing our thoughts, we must take heed that we do not misuse an old language; that we do not narrow our art by becoming specialists in this or that dialect; that we do not lose our proper independence by imitating the peculiar idioms or tricks of expression of any other author or set of authors; and more especially, that we do not waste our strength in futile efforts to invent a new tongue. We must have a care not only as to what we say, but as to how we say it. The individual can no longer be lost in his art.

Modern architecture thus allies itself more closely with humanity than ever; it must appeal, not to the senses alone, but to the mind and heart. Indeed, so saturated is it with humanity that we apply to it moral terms: we say that it is sincere or insincere, true or false, self-denying or self-indulgent, proud or debased. Or we speak of it as a thing of the intellect: it is learned or ignorant, profound or superficial, closely-reasoned and logical or shallow and discursive. Such should be the modern decorative arts, according to the high standard set up by the new culture. In this way, apparently, we are to create an art of the twentieth century. It is evidently not to continue a mere art of correct revivals, now of this or now of that school, according to an inexplicable fashion. Beneath these superficial excitements there is growing this new sense of responsibility as to the real duties of art.

Thus, in building a modern church, the problem is not satisfied by accommodating a given number of worshipers for a given cost, with due regard for the decent setting forth of given rites in an edifice which is merely an accurate quotation from a given style, a correct reproduction of forms recognized by antiquarians as peculiar to a certain distinctive era of art. It is no longer sufficient that it is good Romanesque or good Gothic of any age or place. This is practical archæology, perhaps, but not architecture. The matter of accommodation, cost, and rites being the same, the question is, first, as to the most suitable structure according to modern methods, and, second, as to the most available material; then, what forms are best suited to give this structure and material the most honest and elegant expression possible under the circumstances, adapting these forms to the local conditions of fenestration, exit and entrance, aspect and surroundings. The artist seeks not to invent new forms to meet these conditions; they will come soon enough if needed. There is a venerable and inexhaustible language of old forms; there are innumerable traditionary details, developed out of the experience of mankind in former ages; there are devices of construction developed into shapes associated with the triumphs and trials of Christianity everywhere. With the fullness of this language he utters his thought completely, having in mind only the fairest and aptest expression of his idea. To these processes there are essential, as we have discovered, not learning and research merely, not inventive skill and genius merely, not poetic feeling and fine sympathies merely, but all these combined, together with the usual technical qualities which must form a necessary part of the equipment of the architect. The result must inevitably be, according to the genius, intelligence, and inspiration of the designer, a work of art, — not a correct historical reproduction, but essentially a thing unknown before, a veritable contribution to the pleasure and profit of mankind, a step onward. It is of course dependent upon the learning or skill of the artist whether, in using the old forms of expression, he avoids incongruities; and, while it is not of the least consequence whether he commits anachronisms or not, he must see to it that they are not offensive. He may put Greek and Gothic together if he can, or inform his Gothic with Greek principles, which is better, but it is necessary to the perfection of his expression that all the details shall be reconciled one to another and made one whole.

The decorative arts, from the highest to the lowest, are decorative in that they are fitted for a fixed place, and in that place related, in either subordination or command, to the effect of other pieces of art. "All the greatest art in the world," says Ruskin, "is fitted for a place and subordinated to a purpose. There is no existing highest-order art but is decorative. The best sculpture yet produced has been the decoration of a temple front; the best painting, the decoration of a room. Raphael's best doing is merely the wall-coloring of a suite of apartments in the Vatican, and his cartoons were made for tapestries; Correggio's best doing is the decoration of two small church cupolas at Parma; Michael Angelo's, of a ceiling in the Pope's private chapel; Tintoret's, of a ceiling and side wall belonging to a charitable society at Venice; while Titian and Veronese threw out their noblest thoughts not even on the inside, but on the outside, of the common brick-and-plaster walls of Venice." So, also, with the minor decorative arts. Their essential condition of existence is their subordination to a purpose, and therefore the modern standard requires in their design complicated processes of development, similar, though of course in a less absolute degree, to those by which, as we have seen, the most monumental and important results are to be reached.

In the completion of a room for use by the application of color, of fabrics, and of cabinet-work, it would be easy to prove that a perfect result, or rather a result of perfect fitness, the ideal, is not obtained by masquerading in a foreign dress, or by adopting a prevailing fashion of forms or tints, or by any arbitrary inclusions or exclusions whatsoever, for there can be no such thing as fashion in any form of art: a beautiful thing is an eternal joy and blessing; it cannot be affected by the miserable caprices of any combination of upholsterers. It can only be secured by a study of the peculiar needs and uses of the room, its aspect, its shape, and its surroundings; by the discovery of the key of color necessary to the case; by the survey of available precedents for *motifs* and suggestions of form; by the conscientious and intelligent rejection of every fancy which, however dear to us, however fashionable, however picturesque or original or graceful, is not essential to the realization of this ideal. We know of innumerable rooms, decorated in innumerable ways, by innumerable devices, under all degrees and varieties of civilizations, ancient and modern, and according to all conditions of living. These are importunate in suggesting ideas to the modern designer. Without the exercise of the virtue of self-denial he is at the mercy of these thronging fancies, and becomes a mere superficial eclectic. This virtue must be a leading characteristic of the new discipline which we are approaching, both in the greater and lesser arts of decoration. The obvious necessity of exercising it, if we would create works of art, is another proof of the intense self-consciousness which we must inject into our work.

We cannot decorate a simple panel in these modern days in any spirit but that of self-consciousness. If this takes the form of complacency in our own skill or knack, confidence in tricks of color or form which we have picked up, imitations of what has constituted other people's success, we can have no real success of our own. If the self-consciousness is conscientious; if it rejects the temptations of its own genius and knowledge; if it considers first the function of this especial panel, its position and surroundings, treating it according to the natural capacity of the material, — if of metal, adjusting the form of the decoration so that it may be beaten, chiseled, engraved, or cast into shape; if of clay or plaster, so that the form may be developed by modeling; if of wood, so that it may be carved or painted, — and whether the composition is executed in form or color according to these conditions, if this form or color is kept properly subordinate to the rest of the composition, and is content simply to illustrate or decorate the function of the panel as an essential part of a greater whole, we may hope to create a work of art. But we do not undertake to decorate this panel without an unconscious survey of the whole historical field; how have similar conditions been met in previous conditions of life? An Egyptian would have formulated his work according to his religion, and filled his panel with a vertical composition of reeds and lotus flowers, dead with straightness, rigid, precise, hieratic. A Greek would have contented himself with a wild honeysuckle, but would have extracted from it the very essence of beauty, grave, sweet, corrected, and chastened to the last limit of refined expression. A Roman would have chosen the acanthus and the olive, and would have given to them exuberance, vigor, sensuousness, abundance of life and motion, pride, and vainglory. A monastic designer of the twelfth century, taking up the dim traditions of the ornamentation of the later Roman Empire, would have conventionalized the common leaves and flowers of the wayside, and forced them to grow with formal symmetry within the boundaries of the panel. A lay architect of the fourteenth century would have given a consummate image of what such leaves and flowers should be if they had been created for the sake of his panel; their shapes and their motions would have been adjusted to the form of his panel, conventionalized and crowded.

A century later, he would have crumpled, twisted, and undercut the leaves with dangerous perfection of craftsmanship, and they would have wandered wanton outside the limits of the panel; strange animals would have been seen chasing one another among the leafage. An architect of the Renaissance would have remembered the Roman work; but the Roman acanthus and olive, under his hands, would have been quickened and refined with new detail, new motion, finer inspiration and invention. They would have received a new impulse of life, a new creation, a finer and more delicate spirit. But the art would still have been pagan art; not exuberant and ostentatious, but subdued to a strict relationship with the borders of the panel, observant of the centre line, illustrated with conceits of vases, medallions, birds, masks, animals, boys, garlands, and pendants, all obedient to laws of absolute symmetry. For it was the era of the renaissance of learning, the era of *concetti* in literature as well as art. The decorator of the Elizabethan era would have frankly left nature, and covered his panel with armorial bearings and grotesque emblazonments, with accessories of strapwork curled and slashed capriciously. The Saracen would have filled it with his arabesque tangles and pious texts. The Japanese, following immemorial traditions of art, perfected by successive generations working loyally, consummate interpreters of natural forms, would have disregarded any considerations of symmetry, and projected into the field of the panel a spray of natural leafage from some accidental point in the boundary, cutting across a background of irregular horizontal or zigzag bars; a quick flight of birds would stretch their wings across the disk of a white moon, or a stork would stand contemplative upon one leg in the midst of his water reeds, with the sacred Fusiyama in the distance, barred with its conventional clouds; and yet the composition would be suited to no other shape or size than that of the long panel for which it was composed.

In the presence of all these crowding images, the modern designer stands asking, "Which shall I choose, what shall I reject, and why?" They are all his; they are his rightful inheritance, the legitimate language of his art. He not only has all the beautiful things in nature at his command, but he also knows how they have been used by his predecessors; how they have been interpreted and transformed in the service of humanity; how they have been sanctified by old religions, conventionalized and revitalized according to the knowledge, the inspiration, the needs, the opportunities, the emotions of mankind. They have become an expression of humanity, and thus, as we have said, a language of art.

Mr. Ruskin, in a lecture at the Kensington Museum, asserted, with his usual dogmatic force and confidence, "that no great school of art ever yet existed which had not for primal aim the representation of some natural fact as truly as possible." Accordingly, he directed his disciples to the minute study of leaf and flower, grasses and pebbles, shells and mosses. He told them to look into the rock for its crystals, and to look up at the sky for its clouds; to draw them all with delicate care, to carve or paint them with absolute fidelity : for by such processes alone could the secret of decorative art be revealed.

All this experience doubtless is excellent, and to a degree indispensable. But how this drawing and carving have been done by our predecessors ; how they have interpreted nature according to all the moods and emotions of the human soul, and under all the conditions of life; how they have made it a part of the history of mankind, conventionalized it, in fact, for the uses of art, — this is no less important. The artists who practice design and the theorists who dream of it naturally disagree. What! cry the latter, must we go to art when we have infinite nature all around us? When the clover and the daisy grow in the clod beneath our feet; when the sagittaria, with its pointed leaves, the water-cress, and the long reeds wave by the river's brim, and the white lily floats upon its bosom; when the oak leaf and the acorn help to form the shade in which we repose, — must we go afar to learn how these things were carved by forgotten hands upon the capitals and corbels, in the spandrels and panels and friezes, of sacred buildings, six hundred years ago; or to discover in what way they made beautiful the oaken screens and cabinets in the châteaux of the sixteenth century; or how they were beaten and twisted out of ductile iron in the balconies of Venice, or moulded, baked, and colored in the potteries of Palissy and of Sèvres; how they were painted upon the fans or cast on the bronze vases of Japan? On the other hand, the artist says, What are we to do with our heritage of forms? Are we to leave them to the antiquaries to label and classify and set up in museums, or are we to abandon them to quacks and pretenders, the spendthrifts of art, to be worn by them as savages wear the costumes of civilization? In any event, they cannot be forgotten. Every day they are made more accessible. The instinct of mankind is to use them, and we must see to it that they are used in a manner consistent with the dignity of art, with far-reaching research, but with self-control, self-denial, and conscience.

Thus there are two great books of reference for the artist: the book of nature and the book of art, that is, the book of the interpretation of nature by mankind. If we could close the latter and forget it, and if nature were our only resource, the best of us would perhaps become pre-Raphaelite, and we should peep and botanize in a manner commendable to this great prophet. Much of a certain class of errors might be obliterated from modern art; but our imagination, untrained, undisciplined, without food of immemorial experience, would run into unreasonable excesses. The opportunity and the desire to ornament would not be less, but the available resources would be infinitely impoverished. Our observation of nature would doubtless become quickened, but theory would begin its work of transformation or caricature, and the element of conscience in art, as there would be less occasion to exercise it, would be deadened, if not destroyed. The decorator would soon perceive that the natural form could not be sculptured upon his capital, or painted upon his ceiling, or woven in his fabric, or burned into his porcelain, for a thousand obvious reasons, without undergoing some process of transformation. The work of conventionalizing these forms would at once begin; but, in the absence of instruction and inspiration from all precedent art, it would develop slowly, painfully, with barbarous imperfections and childish crudities. Our art would be a strange mixture: there would be, on the one hand, an absolute fidelity to natural forms, interpreted with the skill which would result from concentration of thought; and, on the other, a more prevalent element of barbarous and illiterate invention, covering the surfaces of things with thoughtless repetitions of detail, as on an Indian paddle. We should be relieved from our embarrassments of precedent, indeed, but we should suffer from a new and greater embarrassment of poverty. The embarrassments of our wealth we are now learning to correct by cultivating the ennobling qualities of self-denial and conscientiousness. The embarrassments of poverty could only engender an overworking, and consequently a debasement, of the powers of imagination. Man, with an infinity of thought to express,—for no fate but death could stop the activity of the mind,— would have no competent language with which to express it. He could only utter inarticulate cries, like a child.

Therefore to say that nature is the only fountain of art is incorrect. Ruskin, illustrating this principle, said: "If the designer of furniture, of cups and vases, of dress patterns and the like, exercises himself continually in the imitation of natural form in some leading division of his work, then, holding by this stem of life, he may pass down into all kinds of merely geometrical or formal design with perfect safety and with noble results. . . . But once quit hold of this living stem, and set yourself to the designing of ornamentation, either in the ignorant play of your own heartless fancy, as the Indian does, or according to received application of heartless laws, as the modern European does, and there is but one word for you,—death; death of every healthy faculty and of every noble intelligence;

incapacity of understanding one great work that man has ever done, or of doing anything that it shall be helpful for him to behold."[1] There is much more of this very beautiful language, but when we get away from its spell and return to facts it seems as if we had been listening to a sort of pantheistic hymn. To go to nature for refreshment and inspiration is always wise; but there is refreshment and inspiration also in the works of man. After God had made the green things of earth and all the animals, the creeping, the flying and swimming creatures, he made man, and endowed him with faculties to appreciate, enjoy, and command the rest of the creation. The result was that man immediately began a creation of his own, — a creation of the second order. His materials were not chaos and darkness, but light and nature. The result of this secondary creation is art. To us of the nineteenth century, for whom have been preserved most of the productions of this secondary creation, not by dim tradition but by scientific researches above and beneath the ground far and near, accurately collated, analyzed, and published, — to us, richly endowed as none of our predecessors have been (for literature has

[1] Ruskin, *The Two Paths*, pp. 46, 47.

CONSCIENCE IN ARCHITECTURE. 131

only discovered the true art of Greece and Syria, of Japan and India, for example, within the last forty years), this secondary creation stands as the image of the primary creation in the human mind; and the human mind, doubtless, is the masterpiece of the Supreme Creator. By this agency nature has undergone wonderful transformations; and although the water-lily of Egypt, the acanthus and honeysuckle of Attica, the olive and laurel of Rome, the trefoil, the ivy, the oak, of the Christian builders, the inexhaustible flora of later times, and all the animal creation, from man to insects, by the processes of art have taken new shapes, — although they have been often modeled in " the light that never was on sea or land," — it is not wise to stigmatize these " old things made new " as the product of heartless laws, and as a conspiracy against nature. There is, in fact, as much nature in the minds, which have thus idealized, conventionalized, and changed natural forms, as there is in the natural forms themselves; and these minds and all the forms of art in which their thoughts have been embodied can no more be neglected by the modern designer than can the primary creation itself.

When the gentle Perdita declines to offer

to her father's guests carnations and streaked gillyflowers, because they are "nature's bastards," — the products of

> ... "an art which, in their piedness, shares
> With great creating Nature," —

the wise Polixines replies: —

> "Yet nature is made better by no mean,
> But nature makes that mean : so, o'er that art,
> Which, as you say, adds to nature, is an art
> That nature makes.
> . . . .
> This is an art
> Which does mend nature, — change it rather : but
> The art itself is nature."

This is the thought which I would enforce : Our present conditions of life must give to art in all its forms certain distinctive characteristics. These conditions require the establishment of principles, and not forms, as standards of excellent work. They make forms the language and not the end of art; and they inculcate the enlargement and enrichment of this language by the study of nature and of all the antecedent arts, to the end that we may express our thought in art, as we would in literature, with an elegance, precision, and completeness commensurate with our larger opportunities, our richer inheritance, and our greater resources. Modern design, especially in architecture, has hitherto concerned itself with the parts of speech, and given us exercises in grammar. Now we are prepared to give to art its larger and truer function; to appeal to the sympathies, not only of grammarians and archæologists, of pedants and architects, but of all the people, by speaking to them, not with a dead language, but with a living language of form, which should interest, instruct, and delight them; by appealing not only to their taste, but to their intellect and heart in terms which belong to our reminiscent but progressive civilization, and which therefore those who choose can understand.

The supreme difficulty which confronts the modern architect is, with the knowledge at his command, to avoid sophistications, pedantries, technicalities. The very finest result of high culture, in architecture as in literature, is to utter thought with simplicity.

HISTORICAL ARCHITECTURE, AND THE INFLUENCE OF THE PERSONAL ELEMENT UPON IT.

THE general inference to be drawn from the foregoing essays is, that modern architecture is fundamentally and irrevocably differentiated from all that has preceded it by the fact that modern architects, working with a great and perplexing amalgam of historical styles, instead of being unconscious and organized servants in the consistent and workmanlike development of a definite style, as their predecessors were in Egypt, Assyria, Greece, Rome, mediæval France and England, and elsewhere, have necessarily become a collection of self-conscious individualities, each responsible to his own personal judgment, artistic feeling, and professional equipment. We have also seen that, among the forces which are now working toward the establishment of a certain unity of effort in this collection of independent individualities, are, first, a systematic education in art as it affects the science of building, and, second, the establishment of principles instead of formulas, æsthetics instead of archæology; and we have learned that chief among these influences is the discovery, in the latter half of the nineteenth century, of Greek lines, which represent the highest point of culture ever reached by man in art. By the revival and application of the principles involved in these lines, it is hoped that the great body of precedents by which we are now embarrassed will be elevated, refined, purified, and corrected for our use in modern work, so that our civilization, at once reminiscent and eagerly advancing, may be adequately expressed in our architecture.

I am persuaded that the curious differences between the conditions under which ancient and modern architecture was and is developed cannot be properly understood until the relation of the architect to architecture, until the effect of his personal equation upon the style of his time, shall have been made clear. I cannot but think that a comparative study of these conditions would not only be historically interesting, but would throw some light upon the nature of the responsibilities of the modern architect in regard to his art.

If the history of architecture were nothing more than a story of the rise, growth, and decay of a series of fashions, and of the subsequent revival of these fashions to gratify the capricious tastes of modern times with novelty, it would have but little more interest or importance than a chronicle of costumes or of manners and customs. But if this history is also the history of the development of the human mind, and if it can be shown that this art has always been curiously sensitive to every essential phase in the progress of humanity, and has taken shape in close sympathy with that progress, the study of it must necessarily involve much more than a consideration of technical or artistic manifestations of structure or taste. The field into which I desire to enter in this investigation is by no means a frequented one; indeed, I hardly know of any literature which has adequately explored it.

Writers of general history collect and analyze traditions, documents, and archives with infinite patience and industry, but the evidences of architecture seem to be inaccessible to them, because those who have written upon this subject have confined their studies rather to its external and apparent forms, as if they

were a development of natural forces, than to
the human interests from which they proceeded; the human wants to which they were
adjusted; the human genius and spirit which
gave to them character and direction. Behind
every evident form in architecture is a human
motive, and a great monument, therefore,
which is a concert of innumerable forms,
must, as I conceive, constitute an invaluable
record of the civilization which produced it.
It is not merely an incident in the history of
architecture, but an incident in the history
of mankind. I am persuaded that the time
is not far distant when it will be possible to
infer from the external evidences contained in
such a work the true genius of the times, the
bases of contemporaneous history. Thus the
Cathedral of Paris, to take a familiar example,
if it were properly analyzed (as it never has
been, but surely will be in due time), would
be found to contain not only all that is essential to know of the spirit of the Middle Ages
in general, of the fall of the monastic orders,
of the decay of feudalism, of the birth of
civil liberty, but in specific detail all the religious, social, and political life of the time;
and this, not so much because it is a great
municipal and ecclesiastical monument, not

because it was deliberately intended to express the history of the times in which it was built, but because it is a work of art, unconsciously expressing that civilization in terms the most exalted and beautiful within the scope of the builders.

We may affirm, therefore, in general, that the qualities of the Gothic or romantic and of the Renaissance or classic civilizations are graphically expressed in their respective arts, and that the characteristic differences between these two arts constitute the most suggestive evidence of the corresponding differences in the conditions of the societies out of which they grew. Referring to architecture in especial, I ask you to consider with me some of these differences, not from a technical point of view, but to ascertain, if possible, what sort of men they were who were more immediately concerned in producing these two forms of art, and what were their relative functions in respect to their work.

When Diocletian laid aside the cares of empire, at the end of the second century, while Christianity was struggling for existence, he retired to Spalatro, on the eastern coast of the Adriatic, where he built a fortified palace covering nine and one half acres. This immense pile of buildings was the most important architectural expression of the Decline of the Roman Empire. If archæology had made a true analysis of this monument before Gibbon's time, that historian would have been furnished with a chronicle even more significant of the true spirit and genius of the later Empire than he was able to deduce from any or all of the voluminous documents which form the basis of his history. But it is more germane to our present purpose to note that the first Christian builders, as the late Mr. Freeman pointed out, discovered in the remains of this vast edifice a principle of construction which gave direction and character to the whole body of Byzantine and early Christian art, and had its final and most unexpected development in the art of the Middle Ages.

This principle of construction was simply the starting of the arch directly from the capitals of the columns without the interposition of the horizontal entablature. At this point in history the horizontal members of the architectural scheme, which all through the history of classic art had formed its most conspicuous and characteristic feature, were practically abandoned, and the way was opened for the growth of the vertical and aspiring features which prevailed throughout christendom until the revival of the classic forms in the fifteenth century.

Mediæval architecture, starting from this basis, was a gradual and steady evolution of structural forms, with but little influence from external traditions, and under exceptional conditions of enthusiastic devotion and religious zeal, made especially effective by an extraordinary unity and concentration of effort. It was saved from feudal barbarism and savagery, first, by the learning of the cloisters, through the era of the round-arched Romanesque, and afterwards, through the pointed arch, ogival, or Gothic era, by the coherence of the lay architects or masonic guilds, which preserved, developed, and transmitted a compact and consistent body of building traditions. This process of evolution, from the twelfth to the fourteenth century, progressed with amazing rapidity: it pushed Gothic art to its highest expression at the end of the thirteenth century, and at the beginning of the fifteenth this art had apparently said all that it had to say; it had become an architecture of *tours de force*, of conceits and grotesques. As its strength failed, it became attenuated and affected, and, at the end, the skill of the craftsman triumphed over the inspiration of the artist. It was, I say, an architecture of structural evolution; and so long as the evolution continued in a healthy condition, it was as free from abnormal examples as the evolution of any animal or vegetable type in natural history. The successive buildings of the style were links in a continuous chain, each one essential to the development of the type. By this series of tentative processes, the art of constructing arched ceilings with small stones, which technically is really the basis of true Gothic art, progressed from the simplest and most timid vault to the most complex and daring, each step in the progress involving changes more or less fundamental in the supporting walls, piers, and buttresses, and hence affecting the entire structural and architectural character of the building.

But it should be understood that a mere evolution of structure cannot in itself constitute a style of architecture. We have in modern times an uninterrupted evolution of structure, which, like the Gothic art, consistently " broadens from precedent to precedent," absorbing into itself all the inventions and discoveries of science in the art of building : but this evolution is in the hands of engineers; it appears in the steady and admirable progress of achievement in works of severe utility, — bridges, aqueducts, sewers, canals, embankments, factories. But this is not art, because, unlike the Gothic evolution, it is not in the hands of artists. The functions of the modern engineer and that of the lay builder of the Middle Ages differ in this important respect: the one develops a theory of construction with a view to obtaining perfect mechanical fitness and stability in the most direct and economic way possible, and with a complete disregard, if not contempt, for æsthetic conditions; the other, in like manner, aims at fitness and stability by the use of scientific and technical methods, but he confers upon his work beauty as well as fitness, grace as well as stability; he aims to charm the eye as well as to satisfy the mind with harmony of proportions and elegance of detail, following in this respect the evident system of the great Creator, who never made anything ugly while making it strong. A Cistercian abbey, produced under the rule of that most rigid of mediæval reformers, St. Bernard, in its austere, unornamented construction, shows how a severe work of engineering may be made a work of art.

A Clunisian abbey, resulting from the more generous discipline of the older order, is no more a work of art because of its superadded luxury of sculpture and splendor of color. The modern engineer gives us prose, the mediæval builder gives us poetry.

If painting and sculpture, and to a certain extent music, are imitations of nature infused with the spirit of man, architecture, which is not an imitation of nature, is a direct growth from the spirit of man; indeed, it is a creation of the second order, and consequently, as I claim, an expression of human character so accurate that all the phases of the essential development of that character in the history of civilization should be read in the successive manifestations of this peculiarly human art.

In the beginning of the period of the Renaissance, the Gothic tree having exhausted itself with blossoms, there was grafted into its enfeebled stock a new shoot, which, with the advancement of learning and the growth of the human mind, gradually but surely changed the character of mediæval forms, until at length they disappeared entirely. Indeed, so completely did this new influence take possession of the human mind that, after hardly a century of transition, the vast poetic monuments of mediæval art became, in the midst of the new civilization, not only barbaric enigmas, but objects of insult and contumely. The complete neglect or fundamental misunderstanding of the principles of Gothic art — a neglect which prevailed from the fifteenth to the middle of the eighteenth centuries, while literature seemed to have hardly a consciousness of the existence of this form of art — constitutes in itself a revelation in the history of the intellect not to be obtained from any other source. Meanwhile the great functions of the church were celebrated in the cathedrals, and generation after generation worshiped within, leaving no trace except in the removal of the delicate tabernacle-work of altar, reredos, screen, and chantry, and the substitution of correct work in the classic taste, without a suspicion of incongruity. Secular buildings were constructed against their solemn walls without a thought of sacrilege. Every marketplace in Europe was overshadowed by the aspiring front of a Gothic façade with its rich traceries, its fretted gables, its buttresses, pinnacles, galleries, niches, statues, and spires; but these great piles for three centuries were as strange and uncouth to those who lived in these hallowed precincts as the Aztec monuments or the Buddhist topes and temples of India are to us. They apparently uttered no word and furnished no thought or pleasure in all this time to prince, priest, or peasant; and so they waited in silence and patience until the slow antiquary of modern times began to discover in them the symbols and handwriting of an earlier civilization, the types of a consistent art, not entirely barbaric or without meaning.

Now what was the nature of the new principle of art which meanwhile was powerful enough to so completely preoccupy the minds of the builders? It was based upon an architectural formula, the classic orders. This formula was the standard by which architecture henceforth was to be measured and corrected.

If mediæval architecture was a system based upon the free and unobstructed development of structural forms, that of the Renaissance was based upon authority and discipline. The classic formula was recognized as the embodiment of the spirit of the antique world, by the revival of which alone, it was thought, redemption from mediæval barbarism and the new birth of civilization were possible. The essence of this formula was an arbitrary system of proportion, composed of isolated columns or attached pilasters supporting horizontal entablatures. The vertical members were composed of bases, shafts, and capitals; the horizontal members, of architraves, friezes, and cornices. Each of these divisions was further subdivided into mouldings; and as all these parts had developed together, and got their shape by mutual adjustment after many trials, as, with the advancement of refinement, artists became curious in respect to the shape and relative disposition of these subordinate parts, the result was finally the highest expression of architectural style known to mankind. Though this delicate and precise mutual adjustment might have been reached in other styles, as in Gothic, it has only been reached in the classic, and the examples of it are the so-called orders of architecture. It is their advantage that they were developed in actual use, — not by theory but by practice, — and that they grew into shape, embodying in their modifications the experience and feeling of successive generations of Greek and Roman architects, concentrated upon the work of purifying and refining the architectural expression of the simplest form of structure known to mankind, — viz., two posts with a lintel or beam laid across their tops. These formulas or orders received in the fifteenth and sixteenth centuries a final revision at the hands of the great Italian masters of the Renaissance, who studied their proportions with the last degree of refinement and recorded them exactly. They conferred upon the coarse Roman orders a degree of perfection rivalling the perfection of the Greek orders. As we have already observed in the beginning of the second essay, the ancient Romans, in taking their forms from the Greeks, had vulgarized them; but the Italian masters nearly restored the purity of the original type. So developed, these classic orders are the key to almost all that has been done in modern architecture to our day.

Now the classic formula, thus revived and rehabilitated, because it represented perfection of proportion, and had no essential relations with structure, which is always developing, could have no career of evolution in the Gothic sense. Perfection cannot be perfected; it can only be adorned. In fact, the essential principle of the architecture of the Renaissance is a scheme of decoration based upon a dogma of proportion.

To our present purposes it is important to observe that the invention and application of ornament to structure, like the Gothic, in process of evolution, is a very different thing from the invention and application of ornament to a rigid formula of proportion, like the classic orders. The consideration of this difference brings us at once to the point to which I would draw your attention.

"Architecture," says M. Veron, "even when considered from the æsthetic point of view, remains so dependent upon geometry, upon mechanics, and upon logic, that it is difficult to discover accurately the share our sentiment and imagination have in it." This is certainly true of Gothic art, because its character rests upon geometry, mechanics, and logic, applied to the harmonious and consistent evolution of structure. But M. Charles Blanc observes of this architecture that "every structural necessity became a pretext for ornament, and the most capricious conceptions were in reality nothing more than contrivances for embellishing the work, forced upon the artist by the inexorable law of gravitation." This is equivalent to saying that Gothic art was an art of construction more or less ornamental. Now ornament is the product of sentiment and imagination, but the invention of ornament in such a service as Gothic art did not call for the highest artistic effort in this domain. The grouping of pinnacles upon buttresses, the filling of apertures with rich tracery, the fretting of sky-lines with crockets and finials, the enrichment of mouldings by conventional billets, beads, chevrons, ropes, and leafage, the decoration of wall surfaces with an embroidery of carved diapers, cusped arches, and canopied niches, — this sort of work was not the labor of illustrious masters, but of nameless trained craftsmen ; nay, even the iconography and symbolism which crowded the porches were part of a hieratic system, to which the men who produced it were in the position rather of humble catechumens or neophytes than of independent artists. Their labors were in no small part didactic, like the labors of writers and printers. In the absence of books and of the ability to read them, the carvers disseminated the doctrines of the church, told stories, imparted ideas, religious and secular, by the works of their hands, in a conventional and childlike language of images, pictures, and symbols, which even the unlearned could understand. Gothic ornament was thus, to a great extent, intended not only to decorate, but to teach.

On the other hand, the composition of ornament for its own sake did not become a leading principle until after the revival of learning, when a new style arose, based upon a rigid formula which was itself perfection, and therefore, as I have intimated, incapable of progression. Under these circumstances, the variety demanded of art by the newly awakened civilization was only obtainable either by distorting or degrading the formula itself, as was actually done in the eras of the Baroque in France, the Rococo in Italy, the Chirrugueresque in Spain, or, when the formula was respected, by overlaying it with ornament not necessarily suggested by the formula, ornament invented and adjusted by the creative spirit of man. This formula also, as, according to the Roman system, it was not materially affected by structure, but applied to it, and, as it was to a large degree inflexible for the expression of special ideas, or of the purpose of the building to which it was applied, depended mainly upon its ornament, which therefore had to be of a character sufficiently imaginative to convey these ideas by emblem, attribute, or allegory, like simile and metaphor in poetry. Thus, unlike the nameless Gothic craftsmen, the famous ornamentists of the Renaissance, not having occasion to teach, were only concerned to beautify and illustrate.

It is evident that this process of decorating without degrading the unalterable classic type could not have been effected by traditions, or by any such anonymous body of trained artisans as kept the Gothic evolution in a workmanlike track. The contingency called into existence the modern architect, with his books and prints and his archæological equipment. The composition and application of ornament demanded a new and higher quality of invention and imagination, because ornamentation had become a far more important function in architecture. The energy and enthusiasm of the intellectual life, newly awakened after the sleep of the Dark Ages, found its natural expression in a finer and more imaginative quality of invention than was possible to the pious craftsmen who, in the service of the church, built their best thoughts into the crockets and gargoyles of the cathedrals, and were blest in their humble self-sacrifice. In short, the Gothic method resulted in an impersonal art; the Renaissance method necessarily led to a personal art.

If Gothic art was the development of principles of construction applied to meet similar requirements, each experimental abbey or cathedral differed from its predecessor in exact proportion to the progress in the art of building vaults with small stones, not in proportion to the genius of the master-builder. The decoration of these structures was not an essential incident in this scheme of general development. No single monastic or lay builder could impress his individuality upon this mighty advancing tide. Its progress was made up of forces far too great to be turned this way or that at his bidding. He was the servant of evolution, and not its master; he made the evolution consistent, grammatical, and beautiful, and in this disinterested service his name and character were lost.

Brunelleschi and Alberti were the first Italian artists who, in the beginning of the fifteenth century, studied the ruins of ancient Rome; they made the earliest essays in the renaissance of classic architecure. Now, if these artists had discovered in these ruins (as the first Christian builders did at Spalatro) a new principle of construction, capable of indefinite expansion in the wider fields of the new civilization, instead of an order of architecture, a more or less inflexible dogma of proportion, and if Francis I. had carried that principle back to France with the spoils of his first Italian conquest, instead of taking with him a crowd of Italian artists, the guardians of the ancient formula, we might have witnessed the phenomenon of another impersonal art, like that of the mediæval period, to the development of which, in the course of time, the genius of the great French Renaissance masters, Philibert de Lorme, Pierre Lescot, and the Mansarts would hardly have been necessary. If, on the other hand, at the end of the Romanesque period in the twelfth century, a certain form and proportion of vaulting and abutment had been invested with such peculiar sanctity that it would have been impious to vary from it henceforth in any essential particular, decoration would have at once assumed the first instead of the subordinate place in the architecture of the thirteenth and fourteenth centuries; and, instead of a series of nameless masons and master-builders, we should have had a catalogue of individualities as illustrious and specific as that which confers its peculiar personal interest on the history of the Renaissance.

Now as to the evidences of the personal

character of Renaissance art, although the classic formula was set up and accepted as absolute authority in the fifteenth century, although it has been used with veneration for nearly five centuries up to the present time, and although every architect since Palladio, designing in the style of the Renaissance, has intended above all things to be correct in his use of this simple type, and to build according to the Italian taste, the result has been, not monotony, not cold and colorless uniformity, but a variety of expression elsewhere unknown in architecture. In studying these variations as they are exhibited in the buildings of the Renaissance up to the present day, we shall find that they are not capricious or accidental; there was one class of variations in France, one in England, one in Germany, one in Spain, one in Holland, one in Italy. These distinctions of class are easily recognizable: they follow natural laws, they interpret national temperament or genius by a visible demonstration. Thus the French Renaissance abounds in elegant variety; it is nearly always refined, delicate in detail, and full of feeling and animation. The English Renaissance is for the most part unimaginative and heavy, feeble in invention, prosaic and cold; but it was not without its single era of great and original development at the hands of Wren and his followers after the London Fire. The German follows the French at a great distance; where it has not been merely imitative, its principal characteristic has been hard correctness and pedantry, purified and enlightened, however, in the beginning of the nineteenth century, by the Greek revival of Schinkel and Klenze. The Spaniard handled the orders with great freedom, and overlaid them with a sort of barbaric and ungrammatical splendor, which is almost Oriental in its profusion and color. The Dutch variation was homely and honest, like the people. But in Italy, the natural birthplace of the Renaissance, Vignola, Serlio, Palladio, Bramanti, Scamozzi, first taught the world how true artists, holding alike to a rigid formula, could invest it with purity and delicacy; how they could be correct without dullness, precise without pedantry, poetic without license. These were the earlier Italian developments; the later masters, betrayed by the contortions of Borromini, the undisciplined inventions of Bernini, the coarse magnificence of Michael Angelo, lost their purity in profusion, and vulgarized the type in their passion for grandeur. Still the Italians were always the true masters of the Renaissance, and the rest of Europe went to school to them. But we have seen that though the architects of France, England, and Germany tried to imitate the Italian manner, they were always French, English, and German; the national spirit betrayed them. All the forces of art were united in the desire to make the revised classic forms cosmopolitan. They studied the same authorities, learned by heart the same formulas of proportion, copied the same monuments, but failed in their efforts to make a universal architecture. Indeed, there was not only a national impress left upon their works, but no two architects of the same nation could produce similar work. Each interpretation had a character of individuality; the phenomenon remains true to this day.

A late writer very happily said that "architecture tells us as much of Greece as Homer did, and of the Middle Ages more than has been expressed in literature. Yet," he adds, "it has been silent since the thirteenth century." This latter proposition is after the dogmatic fashion prevalent in modern English criticism, following Mr. Ruskin, the most eloquent, the most fascinating, the most illogical, of all writers upon architecture. It is equivalent to saying that the personal equation in modern architecture, rendered necessary by the adoption of a formula as the basis of design, has prevented the art from exercising its natural function as an evidence of contemporary history, as it did when design was confined to the development of a theory of construction, in the series of Greek temples and mediæval cathedrals. The fact is that architecture, whether personal or impersonal, cannot be forced away from this function, even by the most determined exercise of the modern privilege of masquerading in the trappings of old forms of art. Indeed, it is sufficiently evident that this very personal character imposed upon architecture by the conditions of the Renaissance, in rendering it more sensitive to impressions of environment, has made it a more accurate exponent of the quality of the civilization which produced it than has been the case with any form of impersonal art, the characteristics of which must be largely due to its own internal forces and structural conditions. While the development of structure depends upon science, that of ornament is poetic, and, like all poetry, must be an expression of contemporaneous civilization, as the poetry of Homer, of Virgil, of Horace, of

Dante, of Chaucer or Milton, of Molière or Béranger, of Byron, Wordsworth, Tennyson or Whitman, indicates respectively a condition in the progress of our race. Each is a product of its own especial era, and would have been impossible in any other. It is therefore not so much its structural as its decorative element which makes the metropolitan cathedral a reflection of the mediæval life which swarmed about the market-place under its shadows, or entered its porches with banners. But if we survey the progress of mediæval French architecture from the time of Philip Augustus in 1180 to that of Charles VII. in 1422, — from St. Bernard of Clairvaux to Joan of Arc, though in this era more than sixty cathedrals of the first class were built, involving an expenditure of more than $500,000,000 in our money (an activity in building operations hardly surpassed in, history), — we shall find that no monarch and no court was able to impress upon it any characteristic which would enable us to distinguish any one phase of its progress as the style of Louis VIII. the Lion, of Louis IX. the Saint, of Philip III. the Hardy, of Philip IV. the Fair, of Charles V., VI., or VII. There are no such styles; the art of these reigns was an evolution of forces, and, as such, it was an impersonal art. No phase of this evolution is identified by the name of a sovereign, even as a matter of convenience in nomenclature, for by its nature it could not submit to the caprices of any court.

On the other hand, when architecture had ceased to be a structural evolution, and was based upon a formula of pilasters and entablatures, enclosing imposts and arches, decoration or ornament assumed a function until then unknown,[1] and there was a distinctive style corresponding to the character of each reign, often contrasting with curious abruptness. Thus we recognize a style of Francis I., free, elastic, poetic, romantic; of Henry IV., bold, coarse, grotesque; of Louis XIV., full of grandeur without, of pomp and splendor within (Viollet-le-Duc has called it " the new Renaissance "); of Louis XV., frivolous, licentious, ostentatious, rococo; of Louis XVI., decent, orderly, pure to prudishness; of the First Empire, a theatrical display of imperial

---

[1] "Strip a chapel of the fifteenth century of its ornamental adjuncts," says Symonds in his *Renaissance in Italy*, "and an uninteresting shell is left. What, for instance, would the façades of the Certosa and the Cappella Colleoni be without their sculptured and inlaid marbles?" *The Fine Arts*, p. 80.

Roman properties, meagre and tarnished; of the Second Empire, elegant but profuse and luxurious, drawing its *motifs* from every historical source, but harmonizing them with a fastidious, academical spirit, not unconscious of the purifying and ennobling influence of the newly discovered Greek principles of art, — a style of to-day, too near to us to be recognized now in just perspective, but for which our French contemporaries will be held responsible in the next century. So there is a style of Elizabeth, of the Stuarts and Tudors, of Mary, of Anne, of the Georges, and of Victoria. So, also, in Spain, the enthusiasm, the efflorescent splendor and pride, of Ferdinand and Isabella and of Charles I. are clearly expressed in the joyous and sunny exuberance of the architecture of their times. But in the time of Philip II. there was promptly substituted for this brilliant manifestation a cold, academical formalism, in full sympathy with his iron rule.

This astonishing display of sympathy is rendered possible by the fact that the architecture of these eras must necessarily be saturated with the subjective personality of the architects, in order to its proper adjustment to the spirit of the times. In fact, the composition of ornament, like the composition of poetry, cannot be evolved from theory alone; neither can exist in definite shape save by virtue of the individual character, attributes, and mental equipment of the author, as they are affected by the era in which he lives. It is his genius which confers upon it all its specific quality. It is his knowledge, his training, his convictions, his taste, which so adjust ornament to a formula of proportion as to confer upon it especial character and interest. It is true that a large part of Renaissance ornament, as of all architectural ornament, is conventional. The enrichment of the orders is obtained not only by the imaginative arts of sculpture and painting, which must necessarily be infused with personal genius like a poem, but by ornament of convention, which is common property, *i. e.* certain types of ornament applied by general acceptance to the decoration of the mouldings and other details of the orders; but it is also obtained by a certain hardihood in varying their proportions. But the use of conventional ornament implies choice, discretion on the part of the architect, and these are only personal qualities; and the variation of the proportions is an appeal to his powers of invention, corrected by his academic training, by his knowledge of precedent, and by his artistic feeling. If an ignorant man plays with proportions, the result is inevitably disgrace and vulgarity. Without technical training, he can produce successful results only by an abnegation of himself, and is only acceptable when he follows the classic formula with unimaginative fidelity. It is only the scholar who can handle his orders with freedom. The advancement of archæological learning, the accumulation and accessibility of architectural precedent, in making it impracticable to the modern architect to build better than he knows, have inevitably forced him into an increasing degree of self-consciousness in his work, and, in depriving it of the grace of innocence and naïveté, have made it a more sensitive index of his individuality.

It is evident, therefore, that the present and future architecture of all civilized nations must, unlike that of all primitive peoples, express, not so much the natural man as the artificial or conscious man, — the man of learning and of resources which have been gathered together out of the past, and distilled in the alembic of his own personality. The picturesque and romantic days of an impersonal architecture will return no more; but the art, in the future as in the past, whether personal or impersonal, must continue to embody the essential parts of the history of the human intellect.

THE ROYAL CHÂTEAU OF BLOIS,

AN EXAMPLE OF ARCHITECTURAL EVIDENCE IN THE
HISTORY OF CIVILIZATION.

In the Château of Blois, as it exists to-day, are brought together three clear and brilliant architectural expressions, each developed from a distinct era in the history of mankind. These expressions are so characteristic and so connected as to invite a comparative study of them, with a view to showing, by an analysis of this monument, how architecture may be used as an evidence of civilization. In making this attempt, I am not without hope that, though confined within limits too brief for an entirely adequate exposition, I may be able to illustrate the intimate and inevitable connection which must exist between any genuine and unaffected demonstration of architecture and the especial conditions of that phase of the history of humanity which produced it, and of which it must be an exponent. I hope to be able in this way to show that architecture, instead of being an art of pure convention, the full appreciation of which cannot be obtained without a comprehension of certain technical qualities inaccessible to any but trained intelligences, has really taken shape and style coincidently with the progress of mankind; that it is an essential part of that progress, and has been closely dependent on it for all those qualities which give it value in art. It is therefore an art of humanity, and may be read like an open book.

The traditions of this stronghold go back to the time of the Romans. But all constructions previous to the thirteenth century have either been entirely obliterated, or have been buried in the mass of the more modern walls. Of the work of the thirteenth century there remain, on the north corner of the quadrangle, the great mediæval hall of the castle of the counts of Blois, divided in two by a central range of arches, each half of the hall having a barrel vault; and, in the west corner of the pile, the lower parts of the round tower, called du Moulin. All other work of the thirteenth and fourteenth centuries has been destroyed, or imbedded in the foundations and converted into crypts and dungeons.

This early mediæval work has no value

from an architectural point of view. But it contains sufficient evidence that the lives spent by the immediate predecessors of Louis XII. in these rude chambers, with their narrow stairs and tortuous communications, and in these dark groined galleries, must have been coarse, gloomy, and meagre, full of lurking suspicions and baleful secrets, when compared with those led by the later households of royalty, upon whom the bright influence of the revival of learning had been shed through open windows, and whose intelligence had been quickened by the tales of Boccaccio, and refined by the poems of Dante, Petrarch, and Ariosto.

It will be observed, on examining the general plan of the château in its present condition, that it consists of three wings, so disposed with their subordinate pavilions as nearly to inclose a *cour d'honneur* in the form of a distorted quadrangle, this irregularity being the most conspicuous and characteristic bequest left to the new structures by the old mediæval castle. The northeast side of this court was built by Louis XII. at the end of the fifteenth century; the northwest side, by Francis I. in the middle of the sixteenth century; and the southwest side, by François

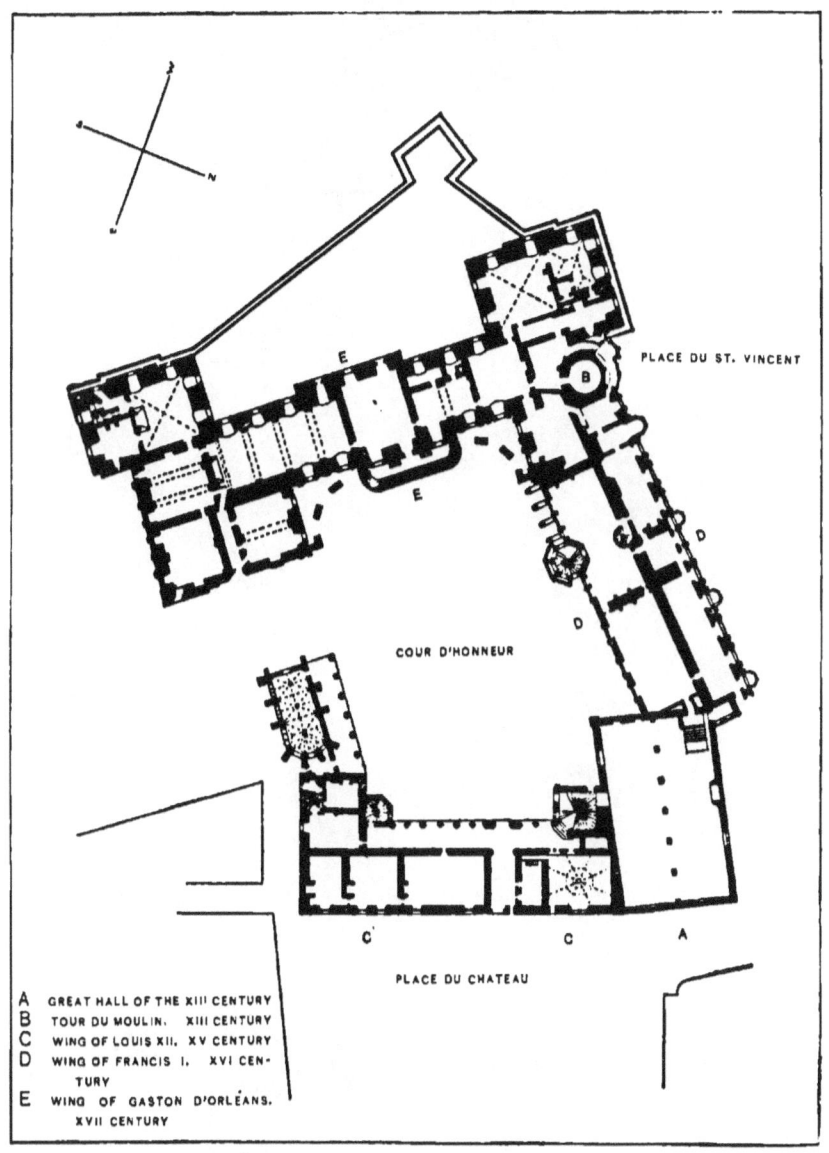

A GREAT HALL OF THE XIII CENTURY
B TOUR DU MOULIN. XIII CENTURY
C WING OF LOUIS XII. XV CENTURY
D WING OF FRANCIS I. XVI CENTURY
E WING OF GASTON D'ORLÉANS. XVII CENTURY

CHÂTEAU OF BLOIS: GENERAL PLAN

Mansart, who in 1635 began to build this wing as the palace of Gaston d'Orléans, after having overthrown the buildings of the fifteenth century which had occupied that part of the site. The southeast side is partially formed by the chapel of St. Calais, which is a part of the wing of Louis XII.; the constructions which formerly completed the inclosure on this side, and which belonged partly to Louis XII. and partly to Gaston d'Orléans, having been lately swept away.

The whole castle, having been nearly ruined by neglect, pillage, and revolutions, has been restored in the interest of monumental history for the French government by M. Duban, who seems to have performed his difficult and exacting task with prudence and archæological exactness. At all events this learned architect has left here three distinct masses of building, in which the capable eye may read much of the history of the fifteenth, sixteenth, and seventeenth centuries.

We present herewith illustrations of the two façades of the wing of Louis XII., the one facing outward on the Place du Château, and the other inward on the court.

The earliest experiments elsewhere in France

with the new style borrowed from Italy, though they already had attracted great attention, were in reality strange and fantastic, entirely destitute of true classic feeling, and not of a character as yet to commend themselves to a monarch so truly French and so intelligent as Louis XII. The Cardinal Amboise was at this time erecting his Château Gaillon, having introduced into France Fra Giocondo as his architect. It is evident that the Italian artist had really little to do with the Cardinal's palace; for it bears every indication of the grotesque travesties to which an organized and disciplined style must submit when executed by men of high imagination, trained only in a free and unrestrained style like the Gothic of that time. The latter in its later domestic forms in France was too beautiful to be readily exchanged for such wild phantasies as those of the Château Gaillon.

Happily, therefore, we find in the new construction at Blois no trace of this spurious Italian influence, though Louis and his father had both been greatly interested in the classics, and had imported Italian artists, if not to control, at least to advise in the work of preparing adequate accommodations for a ceremonious court, the necessity of which was

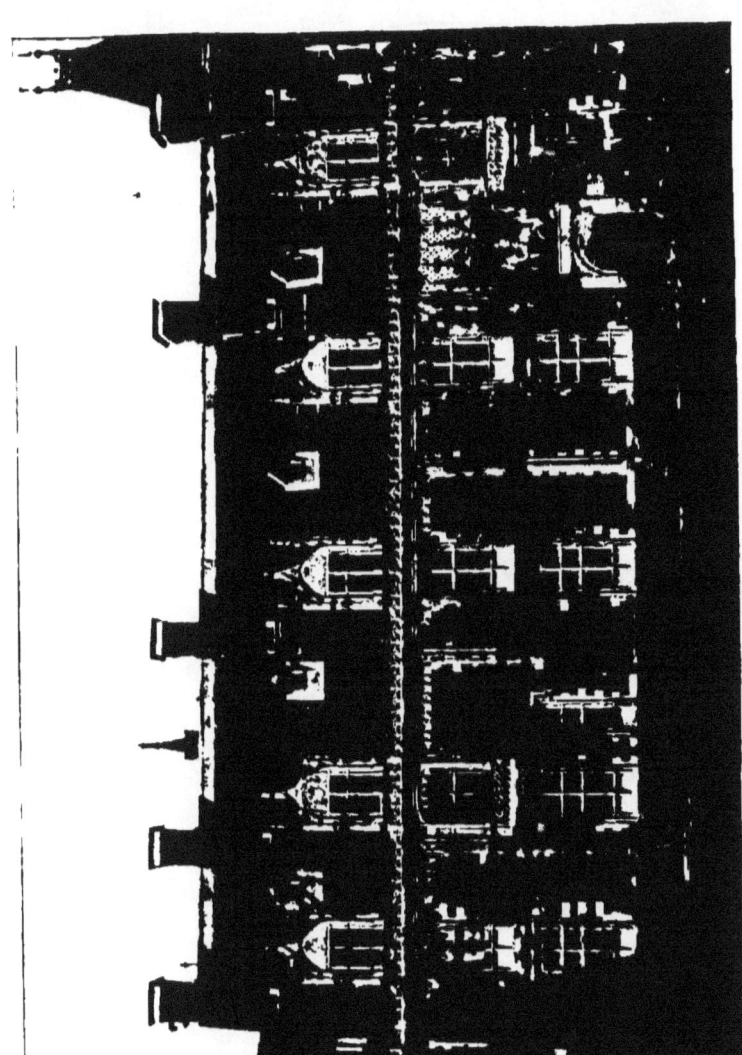

CHATEAU OF BLOIS: THE WING OF LOUIS XII. EXTERIOR FAÇADE.

then for the first time felt in France, and of acclimatizing in that country a style associated with the more advanced civilization of Italy.

The building is in black and red brick, arranged in faint patterns, combined with light stone. It has a high purple slate roof, conspicuous chimneys, beautifully studied dormers, both large and small, and much lead flashing about the dormers and chimneys and on the crests, this metal work being richly decorated in gold and colors with the insignia of Louis and Ann, — the porcupine, the knotted cord, the fleur-de-lis, and the ermine.

The details of this building bear all the evidences of a late development of a style long practiced by a highly gifted race of native builders, — a style which, though it owed its existence and its essential characteristics to ecclesiastical conditions, was clearly undergoing an interesting and highly picturesque process of secularization. The fenestration is adjusted less to satisfy considerations of symmetry than to the most convenient service of the apartments. The general disposition of the mass is stately, formal, temperate, and composed; but in detail it is picturesque and ornate to a very high degree, and very intelligently and thoroughly studied. The moral influence of the renaissance of learning must have been most powerful in being able to supplant with its architectural formulas a native style so rich, so highly organized, and so full of capacity for expression. The wing of Louis XII. represents, indeed, the very highest point reached by the pure domestic architecture of the Gothic period, and indicates that even in those troubled times there existed a clear ideal of tranquil security for private life. Possibly this architecture is its last expression; for it is evident that the style has been enriched, and perhaps to a certain extent sophisticated, by accumulations of ingenious precedents developed from its own conditions. Yet, even in an example so late as this, there is a sufficiency of daring and original design, which is indulged in without straining the resources of the style in the least degree. The great gateway entrance to the *cour d'honneur*, flanked by its rich engaged shafts supporting the conventional Gothic framework and canopy of the famous niche, wherein, against a rich background sown with fleur-de-lis, rides in state the king upon his charger; the little sally-port beside the gate, to provide for which with such naïveté and grace classic art might be challenged in vain; the stone

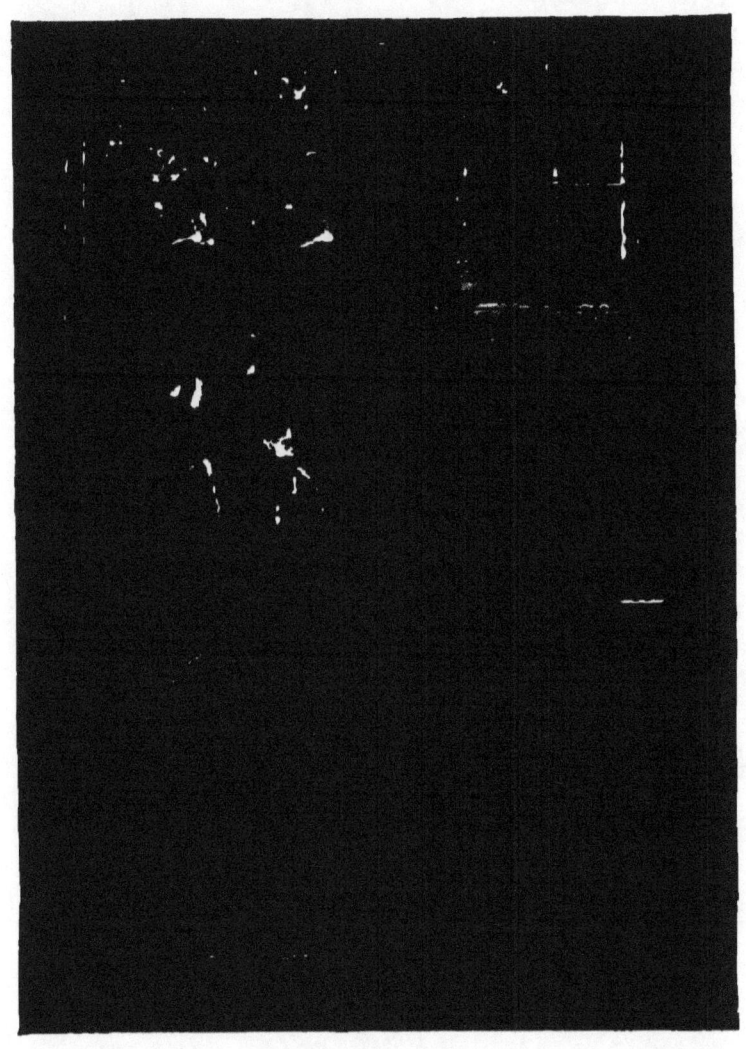

CHATEAU OF BLOIS: THE WING OF LOUIS XII. MAIN ENTRANCE OF EXTERIOR FAÇADE

details of the windows, with their delicate double jamb shafts and their ingenious intersecting mouldings; the beautiful treatment of the balconies; the variety in unity of the great lucarnes; the delicacy of the little dormers; the spacious and elegant hospitality of the great chimneys, — all these things do not seem to point to a decaying style, but rather to one full of life and capacity.

As the modern traveler passes through the archway into the interior court of the château, he finds himself suddenly enveloped by the sixteenth century; the hands are moved back three centuries and a half on the dial of time. Behind him, in the court frontage of the wing of Louis XII., the mediævalism of the fifteenth century whispers its last word.

In some respects this front on the court is a better and more orderly composition than the one which we have been considering. The open portico of the first story, with its flat three-centred arches supported on alternate round and square shafts, all richly decorated, and so characteristic of the later Valois kings, is admirably composed with the five great mullioned windows of this front, and with their noble crown of lucarnes, which, though less rich than those of the outer façade, are far more regular and disciplined in design; the flanking staircase-pavilions forming the interior corners of the court on this side, with their ramping interior groined vaults; the connecting wing on the south, including the little chapel of St. Calais, — all this is truly royal, if not in extent, certainly in refined and ceremonious elegance.

In fact, it seems not at all unlikely that the Italian influences, Bramante being then in the midst of his great career, had their effect here in refining the mediæval forms without the use of the classic formulas, which were then a habit of dress not sufficiently acclimated in France, and not sufficiently understood to satisfy the best critical taste of the time.

The irregular and entirely accidental manner in which the masons were permitted to tooth into the brick-work the carved jamb-stones of the windows is a feature perhaps of slight importance in itself, but sufficient to confer upon the design a remarkable degree of animation. This apparent contempt for mere mechanical neatness and regularity of execution for its own sake is a trait of the artistic spirit,— a trait soon destined to become nearly extinct in practice. It certainly had the effect of enhancing the value of the studied parts of

CHATEAU OF BLOIS: THE WING OF LOUIS XII. FAÇADE ON THE COURT OF HONOR

the design, where the workmanlike qualities of fastidious precision, skill, orderly arrangement, superimposition, correct repetition, and the like, had their proper part to play. This custom of mediæval builders was one of the most fertile sources of their success, especially in works of less than monumental importance.

Now, if an architectural student of intelligence and artistic feeling, but without a knowledge of history, should come upon this bit of royal architecture, and should compare it with that of the neighboring wings of the château, does it not seem quite probable that, as he copied and studied its detail, he could by instinctive inference arrive at a very fair opinion of the quality and degree of civilization reached by the times in which and out of which it was built? There was no masquerading then, no dilettante dallying with foreign styles for the sake of variety, no architectural feigning of impressions which were not genuine, but which belonged to other conditions of life; these are the vices or accomplishments of modern art, and were unknown to the times which we are now considering. Everything that he sees here, therefore, is honest, genuine, and unaffected; he may draw inferences without danger of being led astray.

It would be evident to him, in the first place, that this architectural demonstration, as it gives every indication of fluency and ease of expression and has none of the crudeness of experiment, must be the unaffected language of a people inspired by a native aptitude and fondness for art; that this architecture had gradually assumed new qualities and capacities in sympathy with improved conditions of life; that, as it is fundamentally different from any other contemporaneous style, it was essentially a native and not a foreign growth, was thoroughly imbued with the national spirit, and that it was therefore a free art, intelligible to all, and not to the learned only. The student sees before him a work which cannot possibly be sporadic, exceptional, or accidental, but one which could only exist as the latest expression of a living art, indicating by its open and hospitable treatment, and by the delicacy of its detail, that a gentler and more domestic life had begun to take the place of those feudal conditions, out of which an architecture of defense and defiance had been developed,—an architecture of vast frowning and windowless walls, of battlements, barbican towers, and deep moats. The native grace of expression, which in the earlier stages of its growth in profane architecture had a tendency to a certain grimness of humor, to an unrestrained play of grotesquerie, often coarse, but full of life and power, had here begun to submit to a sentiment of decent reserve and order, in harmony with the manners and customs of a higher civilization. A national architecture of this sort, an unaffected growth from local conditions, is not one to yield promptly to the suggestions of passing fashions or tastes; it is too happily adjusted to the conditions of contemporary life to be made the plaything of fortune. If the court, at the time of the building of the wing of Louis XII., found itself more or less preoccupied by Italian classic literature, pictures, statues, and furniture, it was hardly yet prepared to quote the revived classic upon the walls of the king's palace.

Though the mediæval style had been exclusively used for three centuries and a half and was hampered by the limitations of feudal life or preoccupied by religious emotions, the ingenuity and inborn artistic spirit of the French builders had, up to the close of the fifteenth century, when domestic life was becoming more secure, succeeded in developing, out of these mediæval elements, a secular architectural system, which, though it had begun perhaps to show some signs of fatigue, though it had been somewhat strained, perhaps, in the efforts to meet the exactions of a larger and more complicated ideal of living, and, like all styles, in their later stages of development, had lapsed into an over-refinement and effeminacy of lace-like detail, was still very beautiful, instinct with flamboyant life, and capable of reasonable expansion on certain lines. Certainly this work of Louis XII. of Blois was neither moribund nor decrepit.

The student turns to the wing of Francis I., and finds that, though there is a strong consanguinity between the two masses of structure, they are differentiated most emphatically by the prevalence of classic feeling in the latter, while in the wing of Louis XII. there is absolutely no trace of this feeling. No one, he thinks, can study the wing of Louis XII. without seeing that there was no crying need for reform. Some other impulse, therefore, than mere architectural necessity must have caused this sudden divergence of style. Between the completion of the wing of Louis XII. and the beginning of that of Francis I. (really less than fifty years), there must therefore have come to a head some great moral, intellectual or social revolution, which affected not only literature, manners and habits of thought, and made a new civilization, a complete departure from that of mediævalism, but included the arts, and, among the arts, architecture in its scope. If architecture existed for its own sake like an animal or a plant, we should see it like them going through natural and uninterrupted processes of evolution through the centuries, by survival of the fittest, by environment, etc., etc. But it does not exist for its own sake; it is made by man and for man's use and enjoyment. If there is a revolution in his affairs, architecture must be adjusted as soon as possible to fit the conditions resulting from this revolution.

In fact, while the builders of Louis XII. were constructing here an example of the last development of civilized mediævalism, Columbus was discovering a new world in the virgin West; and while, fifty years later, Francis I. was building at the same place the most important monument of the Renaissance of the sixteenth century in France, another discovery was giving to regenerated mankind another world, no less wonderful and significant, which came out of the immemorial East. The best fruits of the first discovery were hardly ready to be gathered until nearly three centuries had elapsed; the fruits of the second were even then ripening. In analyzing the wing of Francis I. at Blois, we shall discover one of its most prolific and most beautiful manifestations.

This consummate building seemed to belong to a period in history when the forces of civilization were undergoing an evolution, leading mankind from darkness to light, from childhood to maturity, from mystery, romance and chivalry to science and investigation, to letters and commerce, if not finally toward the ideal condition of universal peace and the established happiness of the race. Architecture was becoming conscious of itself; it was beginning to lose its old grace of innocence, and to assume a new grace of knowledge. This building expressed a phase of this development. To study it as an example of a new fashion in art is to disregard its far more potent function as an exponent of the history of humanity at the moment of its most important transition.

When the skeptic, inquisitive light of the Renaissance began to penetrate with fuller effulgence into the dark recesses of the sixteenth century, the secular life of the times experienced a very marked enlargement; it was in fact, more than an emancipation from the bonds of mediævalism, it was a re-birth of the mind, and the revised classic forms from Italy at length presented themselves to the awakened intelligences of these children of art, not as representing an ancient style of architecture, to be quoted with respect and fidelity, but as an essential part of an intellectual revelation, and as a means of expressing their glad consciousness of enfranchisement with a far more adequate and copious vocabulary than had been furnished them by the cramped traditions of the cloisters.

If this new life, this sudden dawn of the modern era of civilization, needs a clearer evidence than is supplied by the general history of this bright period of transition, this evidence may be discovered at the château of Blois in the characteristic contrasts between the work of Louis XII. and that of Francis I. The dates, as we have seen, are less than fifty years apart, but these fifty years were far better than "a cycle of Cathay." We can see here, not merely a change of fashion, dictated by the desire for novelty for its own sake, but an architectural demonstration of a great mental revolution, affecting all the springs of human action. It is, in fact, the most astonishing symbol of emancipation ever presented in the language of art, — first, because it was unconscious, unaffected and sincere; and second, because it was no mere perfunctory repetition of the established and venerable formulas of the classic orders of architecture, but a translation of them, without any hesitation of doubt, into a new architectural system, precisely reflecting the spirited and intelligent independence of the French artists. Nothing could be more insubordinate, and at the same time more joyous and triumphant. The architecture of Francis I. had but little in common with the work of the contemporaneous Italian masters, but its insubordination was not the insubordination of ignorance or barbaric license; for, though it disregarded the established types, its exuberance was not disorderly, and its brilliancy was refined and consistent. It was Gothic in its freedom from classic symmetry and restraint; it was classic in its quick and intuitive comprehension of the true spirit of the decorative architecture of the Romans.

These audacious artists of the sixteenth

century, whose names even are unknown to us, hurried along, as it were, by the swift current of those wonderful fifty years of intellectual regeneration, seemed by some instinct of sympathy to understand, better than any of their successors up to the present time, that the Romans had used the Greek orders, not as an expression of construction, as the Greeks did, but of decoration, having no essential relation to construction; that the development which the classic orders had enjoyed from Augustus to Diocletian had been a development in an entirely different direction from that which they would have experienced if the Romans had continued to use the Greek post and lintel as the basis of their system of building, instead of the arch and concrete vault, to which the orders had been applied as a mere decorative envelope. They seemed by intuition to comprehend that the orders had therefore gradually changed from the virginal purity and fastidious delicacy of the Greeks to the magnificent sensual redundance and ostentation of the Romans, reflecting in this way the character of the conquering race; and that as this system of architectural forms, unlike any other system before or since, had indefinitely expanded and taken shape *independently of structure*, it had become, as it were, a language, flexible to the expression of moods and sentiments, and needed no longer to be confined to the comparatively narrow function of decorating or illustrating a method of construction.

Thus a study of the architecture of this work of Francis I. at Blois clearly shows that these French builders accepted, not the conventional restrictions of the classic formulas, but their essential spirit as an organized scheme of ornament. They did not care for the antique traditions because of the association of these traditions with the greatest triumphs of the race, for they had not yet become scholars; but they were glad to use in their own buoyant way the pilasters, panels, and ornaments, the pediments, consoles, and architraves, which were a part of these traditions, and which enabled them to give free expression to their new emotions. They took up the decorative system of the Romans, and, in a spirit of unconscious daring, carried it on through another stage of development. Their old Gothic, with all its flamboyant capacities, with all its delicate embroideries, its moulded and fretted chamfering, its elaborate cuspidations, and crocketed finials, stood "patiently remote from the great tides of life." It could not give utterance to the joy of a secular liberation, the basis of which was the revival, or rather the adaptation of pagan learning and pagan art. In fact, two centuries ago mediæval art had in the cathedrals given expression to its highest aspirations and its most beautiful thoughts, under an impulse which could never occur again. Reminiscences of the age of chivalry, of crusades, and of hieratic denomination, whether because of a certain sentiment of romantic melancholy, mysticism, or ascetic meditation which lingered about them, or because of their technical limitations when expressed in art, could hardly furnish the expression needed in the architectural manifestations of a time when the revival of learning and letters was liberating reason and imagination from the vagueness and poverty in which they had so long been held by the homilies of the fathers, and by the crude speculations of monastic schoolmen.

By referring to the plan, it will be seen that the wing which we are now considering is divided longitudinally by a very heavy wall; the part of the plan lying outside of this wall belongs to the château of the fifteenth century, and its galleries were extensions of the wing of Louis XII., and formed a part of the accommodation provided for the court of that monarch. The part lying on the other side of this wall, toward the cour-d'honneur, belongs to Francis I., by whom the old part of the wing was remodelled and furnished with a new outside façade.

The façade on the interior court shows that the lessons in the classic styles given by the great Italian masters, who were entertained by the court like princes and ambassadors of art, were received by the native artists with intelligence and respect, but were not learned by rote. These lessons were assimilated, but not accepted as laws. We have here a truly Gothic irregularity of fenestration. The windows of the basement have but little vertical coincidence with those of the two superimposed stories. The former are each flanked by short pilasters, supporting a string course, which is a line of demarkation between the inferior order and that of the important system of pilasters which decorate the two stories above, and which have no essential relation with those standing beneath them. The greater order of pilasters is irregularly spaced so as to have a constant relation with the windows; for the intelligent builders

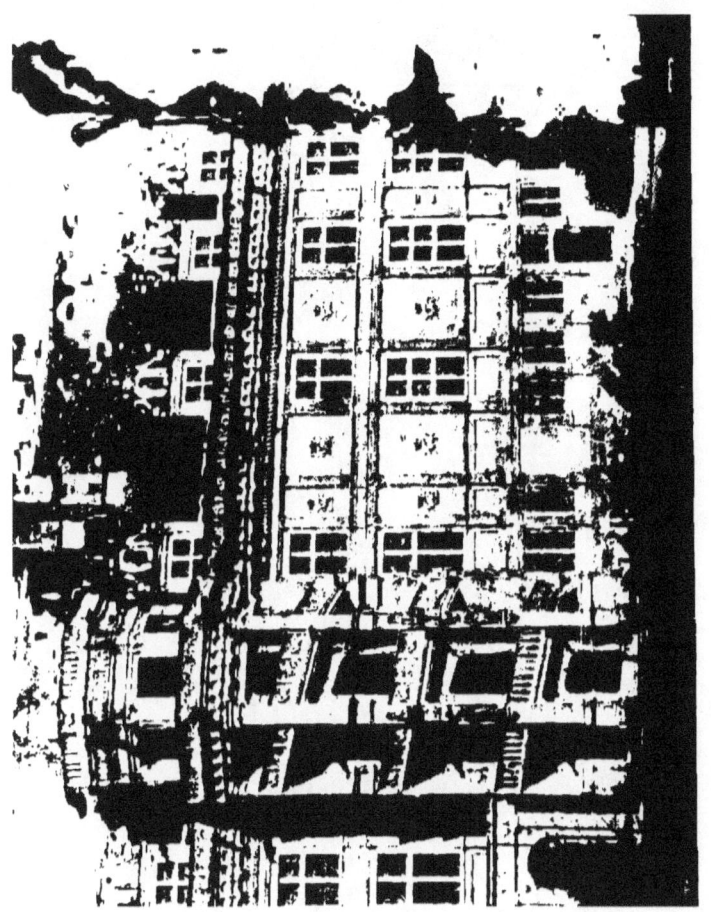

CHATEAU OF BLOIS: THE WING OF FRANCIS I. FAÇADE ON THE COURT

immediately understood that these features, as used by the Romans, had a decorative and not a structural function. There is an entablature or string course between the two principal stories, dividing the pilasters belonging to each into two superimposed orders. The wall face is crowned by a great cornice and balustrade of excessive richness; in this balustrade, for the conventional cuspidated Gothic open work of the corresponding balustrade in the wing of Louis XII., is substituted an open work composed of the royal initials F (crowned), and C for Claude, the queen, entangled with the knotted cord. Above, half hidden, is a range of lucarnes strangely contrasting with their Gothic neighbors, and, though uttered in a pagan tongue, as it were, quite as rich and poetic as those which were expressed in the vernacular. Each is crowned by a Cupid in a canopied niche, which is flanked by flying consoles, ramping from highly embellished pinnacles, and the whole is relieved against an enormous slate roof with massive grouped chimneys. The blank panels of the wall contain each the king's salamander, crowned.

The general effect is as studied and refined, and essentially as classic, as any contemporaneous work in Italy; indeed, much of the decoration in the panels of the pilasters is identical with that of Italy. The differences are not differences of caprice, of ignorance, or carelessness; there is nothing savage, raw, or grotesque in them, such as occurred in the earlier attempts at Gaillon, Ecouen, or even in the Tuileries, and such as distinguish the contemporary attempts at classic by Henry VIII. and those of his immediate successors in England. The correct orders, brilliantly set forth by the scrupulous and finely pointed pencils of the trained Italian gentlemen, finished with the regular formulated details of architraves, frieze, and cornice, and exquisitely embellished with all the delicate resources of their art, instructed but did not dazzle the French masters. The alertness of their minds, the independence of their genius, their instinctive impatience of scholastic restraints, their facility in the invention of ornament, developed from the absolute models set before them a new art, never seen before. They converted the old cornice of defense into a cornice of most delicate beauty. The coarse corbels which supported it were replaced by elegant classic modillions, and the black machicolations between them, through which the defenders of

THE CHÂTEAU OF BLOIS. 187

the castle poured missiles and boiling oil upon the besiegers below, became a series of delicately moulded arches, and each enshrined a shell; beneath was a rich bed of dentils and minor mouldings; above was a powerful Gothic overhanging and undercut moulding, interrupted by gargoyles.

"Peace hath her victories,
No less renown'd than war,"

and architecture crowns the triumphs of the gentler dispensation with beauty.

But yesterday these builders had found in the freedom of their old Gothic a language as copious as they needed, in which to express the spirit of the times; but when the Italians presented them with a new set of motifs and a new theory of design, belonging to the new learning, and a part of the brilliant manifestations of the revived classic art, these unknown and unnamed masters of living style received the pilasters, architraves, friezes, cornices, pediments, and consoles of the Renaissance, and treated them with the same joyous and ingenious liberty as they had been accustomed to use when their language was a language of moulded chamfers, carved labels, and cuspidated tracery. They sang their old songs in the new tongue, but with a grace of expression such as no rigidity of formulas could chill, and an elegant fecundity of imagination such as could be restrained within no classic limitations.

The preservation of the integrity of the classic orders was a function which these free children of art were hardly yet prepared to exercise. The spirit of the civilization of the epoch was one, not of conservatism, but of eager investigation. It was not until the next century, as indicated in the works for Gaston d'Orléans in this very château, that architecture, like all the branches of learning, was willing to submit to authority and to reproduce the classic formulas with respect.

The great glory of this front on the interior court is the famous open octagonal staircase, which detaches itself from a point not far from the centre. This is generally accepted as the most brilliant and effective piece of work of the sixteenth century, but it is also another example of the audacious independence of these French artists. It has no architectural relation with the wall surface in which it is imbedded, except that its four great freestanding buttress piers support a cornice and balustrade, which are a continuation of those of the wall. The power and strong accentu-

CHATEAU OF BLOIS: THE WING OF FRANCIS I. UPPER PART OF THE GRAND STAIRCASE ON THE COURT OF HONOR

ation of this cornice is entirely justified here; nothing less would have been sufficent. This feature of the façade, which might have challenged criticism under other circumstances, thus seems to be quite condoned. The vertical mass of these piers, which, in plan, radiate like buttresses from a common central point, is divided into two sections by horizontal mouldings, marking the line of the third story floor; they are furnished with bases and sculptured capitals. In a niche upon the lower part of the face of each of these piers stands a beautifully executed female figure upon a corbel and beneath a pinnacled canopy, enriched with the most playful fancies of Gothic imagination, expressed in the terms of a new art. All these features are set on lines parallel with the cornice and base. With these the horizontal elements of the design cease. Every other feature ramps with the gentle ascent of the stairs, thus encased. Richly carved balconies, arranged on these sloping lines, are carried from pier to pier, coincident with the sweep of the stairs, stopping against the returns of the piers and near their face. Deeply recessed between each pair of piers, and beneath each of these balconies, is a raking beam supported by richly carved pilasters.

Above the cornice line, this unprecedented composition finishes with an attic order, bearing a decorated entablature, of which the vertical elements are strongly emphasized, and a balustrade. Every panel, large and small, bears the royal insignia, the crowned F and the crowned salamander.

A design so bold as this, so beautiful in its detail, so original, handling the classic formulas with such easy confidence, is certainly an astonishing piece of work to be executed by the children of the carvers who achieved the Gothic building of Louis XII. only fifty years before; but the latter constructed the great vaulted staircase in the corner of the interior court. The skill required to do that, when applied to this new and much more difficult problem, needed to be supplemented by a rare quality of artistic inspiration, sympathy, and excitement to make such an achievement possible. By some purists this staircase has been called "a plaything." In fact it is a useless chef-d'œuvre, which never could have been built in a settled and serious age; but the joyous new birth and enfranchisement of the mind, which were then filling all the corners of Christendom with inspiration and light, could not have their due record in architecture with any less ebullient demonstration.

Turning now to the renowned outer façade of this wing, we recall that it is a mere superficial facing, upon a construction of the previous century, and that the whole composition started from the necessity of establishing a communication, around the outside of the round mediæval Tour du Moulin, between the great mass of buildings forming the southwest wing and that of the northwest wing. This gallery of communication took the form of an envelope of arches in two stories, with a projecting bay, all the arches being separated by engaged columns, whose circular form and vertical lines are carried upward from bottom to top without interruption, after the Romanesque manner. Subsequently these orders were extended for six bays northward, pilasters being substituted for engaged columns, and wide three-centred arches taking the place of the narrow round arches. At a still later period the orders were extended six bays further toward the north, segmental arches taking the place of the three-centred ones, used in the previous section. In this way the whole of the façade was at length enveloped. The roof is supported in the Italian manner by columns free-standing above the pilasters underneath, thus furnishing a continuous protected passage-way connecting the apartments in the roof. Near the centre this continuity is interrupted by a stone dormer in two stories with gable and tall finials.

One great secret of the successful effect of this noble composition lies in the fact that the French builders had not yet quite thrown off their mediæval disregard of absolute symmetries and repetitions. The two orders of the loggia are not arranged according to any mathematical and geometric system of correspondence and contrasts. They are not academical in their dispositions. They have never been set down in text-books as examples for study. The old Gothic spirit of liberty, restrained from license by the instinct of art, prevails from one end to the other of this façade. The loggias grew into place subject in their details to accidents and to adjustments of new forms to old conditions. It would be unprofitable and inartistic to explain these differences of treatment, or to attempt to give a reason for them. This is the business of the archæologist. It is sufficient to say here that these variations of treatment do not result in an effect of careless picturesqueness. The composition is essentially an expression of unity of the highest quality. Its variety does

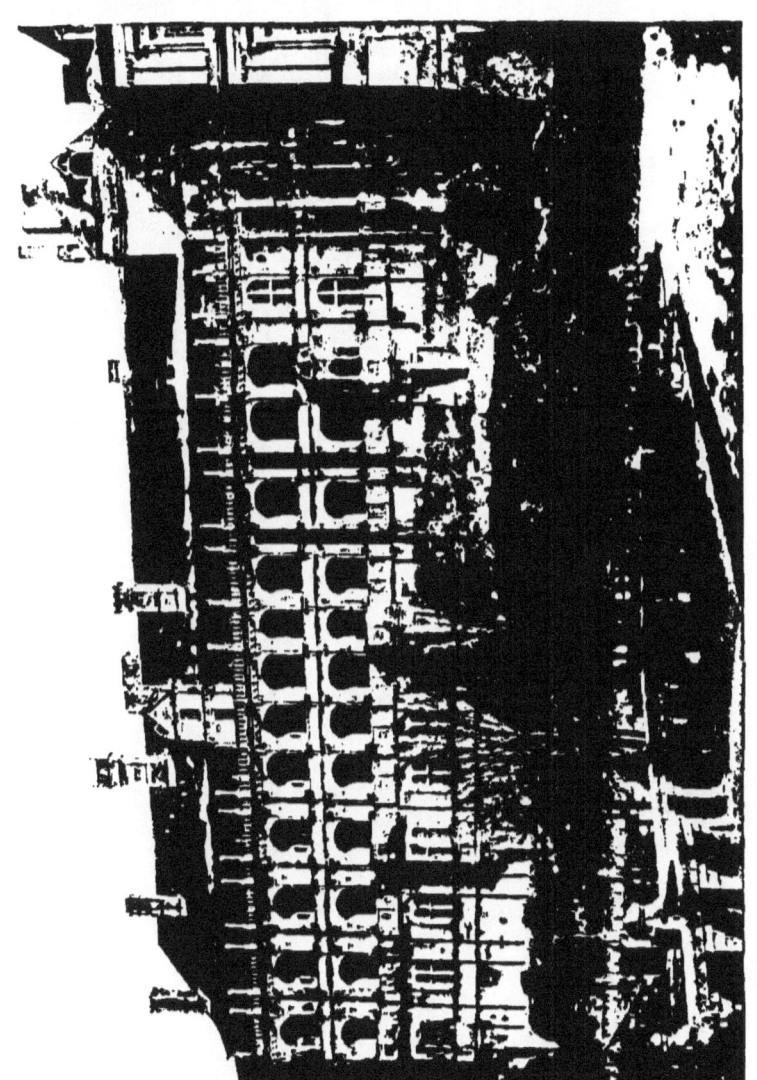

CHATEAU OF BLOIS: THE WING OF FRANCIS I. EXTERIOR FAÇADE

not interfere with, but rather enhances this expression. If it is not strictly and academically classic, if the eye cannot detect by careful study a mathematical balance of parts, with a regular and studied correspondence and recurrence, to bind the whole into formal and ceremonious harmony, the best effect of such harmony is here completely portrayed. Its royal dignity and strength, its noble repose, its processional movement from end to end, present an architectural pageant, unaffected, sincere, and imposing, such as no conditions of civilization elsewhere in the world, at any other period of the progress of mankind, has been able to produce. It is another expression of the deliverance of the Christian world from the bonds of feudalism and its happy entrance into the liberty of modern civilization. It is coincident with the establishment of new colleges, the formation of new libraries, and the increase in the printing of books. It is a reform which did not destroy the old things for the sake of the new; but which adapted them and amalgamated them with apparently unconscious ingenuity, but with the instinct of the true artist-spirit. To laboriously imitate this irregularity, converted into beautiful unity by its subordination to a noble architectural *motif*, would be an affectation which would inevitably betray its author.

The vast basement of the old mediæval structure develops at the northern end a series of round arched coupled windows, each pair collected under a single arch. The ancient substructure is a massive battering wall, growing out of the steep natural slopes of the hill, upon which the mighty fabric stands. Out of this battering surface a series of vertical piers or buttresses is gradually disengaged, each sustaining a pilaster of the superstructure; each pilaster breaking through the horizontal lines of entablature, creates, from foundation to roof, a system of vertical lines, relieving the composition from that predominance of the horizontal *motif* which would have robbed it of its movement, and given to it a ponderous majesty, like that of the great amphitheatres at Rome or Verona. Indeed, the contrast in character between the antique arcades and those of this sixteenth century movement is full of significance when considered in their relations to the history of the mind.

The first and second stories of loggias are connected, the former with the apartments of the queen, and the latter with those of the king. To provide for the proper lodgment of the royal pair, the splendid equipment of the interior chambers must have an adequate palatial expression on the exterior. To this end the series of loggias is broken in four places, unequally distributed, by the projection of two octagonal oriels, near the north end, corbelled out from diagonally placed piers of the root-like substructure, and finishing with carved balconies on the level of the first loggias; the other two are in the form, one of a rich corbelled balcony, and the other, developed from the queen's chamber, in the form of a delicate oriel with a domical roof, thus reaching a higher level than the others. In short, the succession of deep bays is broken by numerous accidental irregularities of detail, not only without affecting the majestic unity and repose of the whole, but conferring upon it a charm of picturesqueness, which relieves it from the evil of monotonous and mechanical uniformity.

The mellowed tones of the masonry on the front are offset by the rich colors bestowed upon the decorations of the deep jambs of the loggias, and the outside world is thus permitted to have an exterior suggestion of the sumptuous tapestries and of the carved and gilded ceilings of the royal chambers within. Here the Duc de Guise and the Cardinal de Lorraine, his brother, were assassinated; the former, in the chamber immediately over the dome of the oriel; the latter, at the door of the chamber of the Tour du Moulin in the queen's apartments beneath.

But whatever dramatic episodes of the dark and sinister politics of the time were here enacted by the accidents of history, the architecture itself seems to hold in solution the spirit of the sixteenth century in France, with all its strongly-contrasted lights and shadows. It is not a theatrical representation of this spirit; it is the spirit itself, genuine, without affectations, and it seems to convey an idea of noble and spacious repose, of majestic security. To the nineteenth century it is alien and strange, like a romance.

In this way the children of the architects of Louis XII. achieved a transition from revery to reality, when, at the command of Francis I., they undertook to build not only at Blois, but at Chambord, at Amboise, at Chenonceaux, at Madrid, and at Fontainebleau. They did not pretend to constitute themselves a guild of the elect. Their inspiration was the genius of the times in which they lived. They were not engaged in an amusing experiment of self-culture with a new set of architectural forms; they were not consciously practicing with a new theory of esthetics; but they were, as artists, sensitive to all the strange movements in the world around them — movements such as never before had stirred the minds and hearts of the nation. This complete development of a new style within less than half a century could not have arisen merely from the creative power or flexible intelligence of the artists, who were the agents in the quick transition, but must have been due to the fact that they were sharing in a new impulse of great moment pervading a whole people, and awakening it into a condition of extraordinary intellectual excitement.

Classic art, under these circumstances, revealed capacities of jubilant expression such as the Italian masters appear never to have suspected. Their renowned Certosa of Pavia, with its boundless opulence of beautiful detail, is perhaps the nearest approach they ever made to such an expression; but it conveys to the mind an idea rather of elegant prodigality, of refined and highly accomplished artifice, kept strictly within the bounds of conventional form, than of the unconscious grace of youth in all its joyousness and freedom. We have already referred to the audacious transformations to which the classic orders submitted on the court front of Francis I. at Blois. In like manner no Italian palace ever exhibited anything like the roofs, dormers, and chimneys of the châteaux of this reign. The design of these features at Blois is free and romantic; yet, when we compare them with the dormers and chimneys of Louis XII. in the adjoining wing, we cannot hesitate for a moment to consider them perfectly fresh, articulate and grammatical expressions of Renaissance feeling, with none of that timidity or gaucherie which generally accompanies a first trial with a new style. In fact, the foreign forms were completely acclimated without conscious effort, and, in the process, they suffered a change, significant of a great transition in the history of humanity.

In the next century, as we may see in the wing of Gaston d'Orléans, which completes the closure of the interior court of Blois, civilization began that modern stage of development, which is expressed in architecture by the spirit of conformity, or respect for the authority of precedents. As the architects became more learned they became academical, serious, exact, pedantic; they lost that graceful and picturesque fearlessness with which the first emancipation of the mind had been joyfully celebrated, and, with the growth of consciousness and the establishment of order, they became correct, discreet, and, in a sense, respectable. The classic ideal, when it began to be better understood, assumed the character of a fetish. It was the age of Corneille and Racine, when the preservation of classic style was of more importance than dramatic freedom or the portraiture of human character. In the same spirit, therefore, the Roman orders of architecture, according to Vignola and Palladio, were loyally reproduced by François Mansart in the last constructions at Blois, presenting, when compared with the almost Rabelasian liberty of the art of Francis I. in the adjoining wing of the château, a contrast full of historic meaning.

This great monument, therefore, comprises within itself a series of evidences of the development of civilization through five centuries of experience; and these evidences are all the more significant, because their close juxtaposition challenges a comparison, which cannot be intelligently followed without revealing all the essential processes and characteristics of this evolution. No lesson is so impressive as an object-lesson. The narrow and gross restrictions of the intellectual life of the thirteenth and fourteenth centuries; its aspirations and its growth in capacity for refinement and elevation in the fifteenth; its sudden and triumphant expansion and liberation in the sixteenth; its firm establishment on the basis of law and order in the seventeenth; all these vital and progressive movements of humanity are recorded on the walls of Blois in characters which can be easily read by any mind open to the appeals of art. If written history is distracted by episodes and accidents, if it substitutes personalities for principles, and is subjected to the moods and prejudices of chroniclers, these unconscious architectural evidences can always be depended upon to correct and to simplify the record, and to reveal the true spirit of the times. Beneath the surface of the detestable politics of the sixteenth century, beneath the treacheries and assassinations with which the pages of the historian are preoccupied, beneath the gossip and scandals of the court, set forth by the inimitable Brantôme, were the great fact of the Renaissance, and the glad beginnings of modern civilization. The monuments do not tell anecdotes of the eras when they were built, or concern themselves with personalities; but they were inspired by the fundamental principles which were directing the progress of mankind.

THE PRESENT STATE OF ARCHITECTURE.

IN a recent number of the London "Athenæum" there appeared a review of a new translation of Arabian poetry. A passage in this review is worth quoting, because, in commenting on the conditions of civilization which affect that form of art called poetry, it uses language equally applicable to that other form of art called architecture, and is thus unconsciously significant of the close analogy existing between the various forms of art in which man has expressed his higher emotions. The passage, with a few verbal changes, mainly in substituting the word "architecture" and its derivatives for the word "poetry" and its derivatives, is as follows :—

"The architecture of a people, who have preserved their natural character and simplicity, and have so far learnt nothing from other civilizations, must always possess a strong fascination. As soon as the period of study and learning arrives we obtain, indeed, forms of architecture and poetry beautiful in themselves, and full of the thoughts and inventions, the spirit and characteristics, of the best works of many nations, but we lose the simplicity, the unaffected naturalness, the fresh outlook upon life and nature, which belong to primitive races. The freshness and sincerity which are exhibited in the architectural works of such races arise from the fact that they formed their style for themselves, with no assistance from other nations, and developed form naturally and out of necessity, with no admixture of preconception derived from books and study. They did not suffer from the difficulties which beset the modern architect; they had no models in other styles *to teach them to affect impressions which they did not feel;* there was no searching after originality with them, since the native and instinctive ideas and forms of art had not been exhausted in their time; and though they spared no pains to attain the utmost degree of artistic finish in their work, they were not ever striving after the discovery of new *motifs*, or rare combinations and tricks of design, to render their work original and interesting."

This quotation may fairly introduce what I have to say on the present conditions of architecture. It is impossible justly to study this theme without constant comparison of the attitude, functions, and methods of the architects who produced what we now recognize as the historical styles, with those of the modern architects, who, far more learned and versatile, far better equipped, are contending with projects of building far more complex, in an atmosphere infinitely less favorable to purely artistic achievement.

The profession of architecture is now reproached because it has failed to establish "a style," because it has not agreed upon a system, because its followers do not move in parallel lines onward towards a consummation of art commensurate with our civilization, in the same way that contemporaneous science has moved towards the development of the electric telegraph, of electric lighting, of the telephone; because at our annual conventions the president of the British or American Institute of Architects is not able, like the president of the National or International Academy of Sciences, to report in his address a definite and orderly progress of achievement.

But architecture is a fine art upon a basis of science; if it were a pure science we could emulate the electrician, the geologist, the political economist, the naturalist, the civil engineer, and report, like them, an annual record of consistent advance in all that relates to questions of construction and practical building methods. Modern architecture, as a fine art, cannot make its annual boast of improvement, for reasons which are well worth investigating. But, on the other hand, Callicrates in his day could have reported to Pericles, if required, a definite progress in the development of the Doric order within any twelvemonth of his career; Apollodorus, in like manner, might have reported to Trajan a corresponding progress in the architectural use of the arch; Anthemius, of Tralles, could easily have described to Justinian a clear advance in domical architecture in any successive half dozen years of the reign; the lay builders of the fourteenth century could have traced distinct stages of growth in all the details of Gothic art from year to year; and Marie de Médicis would not in vain have ordered from Philibert de Lorme a report of annual progress in French Renaissance, nor would Charles II. have been without response if he had thought it worth while to summon Sir Christopher to give account of the conversion of the orders in English hands. In those days styles were visibly unfolding towards perfection; architecture was a living art, developing on lines of consistent and steady progress, without distraction or bias; temple followed temple, church followed church, château followed château in a reasonable and natural growth of architectural forms, confined within practicable limits. The study of the architect was limited to a type which all understood, and there was an orderly, intelligible, and harmonious evolution of styles. The forms in vogue, by means of a series of practical experiments in a succession of structures of the same sort, adapted to a comparatively simple condition of civilization, underwent a process of purification and natural enrichment. They gradually approached, and finally achieved, technical perfection and consistent harmony. At length, when the inherent capacities of the style had been exhausted by use (the human mind declining to rest upon, or, indeed, to recognize, the attainment of perfection, but demanding ever new things, fresh surprises), it was by degrees overlaid and overwrought with invention, it declined with laborious splendor, and, in due time, gave place to a new set of forms, which were introduced in the establishment of a new religion, like that of the Eastern Empire at

Byzantium; in a political conquest, perhaps, like that of the Normans in England; in a moral reform, like that of St. Bernard in France; in an intellectual revival, like that of the Renaissance in Italy; or by the influence of a brilliant court, like that of Elizabeth or Louis XIV., demanding an especial expression of splendor or triumph. And these new forms, in their turn, were developed to completion by the same processes of consecutive experiment in a narrow field of enterprise, and constituted in each case a style, an exponent of manners and customs, with a beginning, a middle, and an end, between two brief eras of doubt, called, in the history of art, eras of transition. It is important to note that these old styles disappeared and these new styles arose, not in the modern manner by caprices of fashion or by theories of æsthetics, but because they followed unconsciously and inevitably corresponding changes in the historic conditions of our race, and were a part of the development of civilization.

Now in the modern architectural chaos there appears to be a notable exception in the work of the French people. In Paris, archæology and the theory of architecture are taught in an official school of fine arts, which is the guardian of the national traditions. In this school the basis of study is the classic formula or dogma of the orders received in the sixteenth century from Italy, and since then adorned and vivified so as to form a great body of national precedent, reflecting the advance and character of French civilization through all its stages. Architecture is in this way officially organized and kept in a steady line of academic development. Thus confined, French genius is not, as elsewhere, exhausted in experiments, or spread thin over fields of enterprise too extensive for a display of effective progress; nor is it distracted by capricious archæological revivals. This concentration of energy expresses itself in a degree of refinement in detail, a degree of clearness and directness of thought, a degree of self-restraint and repose, which are quite unapproached in the practice of any other nation. Under this dispensation technical qualities of design are naturally carried to the highest perfection. Refinement is often pressed to the verge of effeminacy. The highest results obtained under this system are, on the one hand, extreme dignity and repose, as in the Palace of Justice; and, on the other, a poetic and florid, but always a correct, brilliancy, as in the Hôtel de Ville, of Paris, and the New Opera. If the architectural conventionalism which it fosters is sometimes commonplace and monotonous, it is always correct, never illiterate, and often scholarly. If it disciplines individuality of thought, so that the style, in the hands of inferior artists, becomes unduly uniform and uninteresting, it protects common work from the dangerous vagaries of invention, and keeps it pure. Originality is not sought after with the feverish eagerness which must be the prevailing characteristic of work done under a condition of liberty. The school, in fact, is a propaganda of faith in an arbitrary type of art. While it narrows the range of expression, it encourages academic precision, fosters beautiful invention in detail, and leads to a study of ornament far more delicate and precious in its results than is elsewhere possible. As a school for practice and education, it therefore maintains a conspicuous advantage. Viollet-le-Duc, with all his knowledge and all his convictions, eloquently urged in favor of a return to Greek and mediæval methods in design, was unable to create a successful revolt from the national styles as established under this official system. On the whole, modern French Renaissance, with its vast accumulation of *motifs*, resulting from four centuries of constant use in the hands of a naturally inventive and imaginative race, constitutes a language of art, which has become at once homogeneous and copious; but, as this language is scholastic and learned, it is not, like the free vernacular of the Middle Ages, quickened and enlightened by the spirit of the people. If its essential paganism makes it less fit for the expression of romantic, picturesque, or religious thought, and perhaps, by reason of its academical character, less adaptable for domestic purposes, this quality renders it more elastic than any other for monumental and civic uses. It can be gay or grave, profuse or severe, stately or poetic, without straining its resources of expression, and it still continues to reflect the spirit of the times with the same fidelity that has characterized it in all its historic phases from the style of Francis I. to that of Napoleon III.; yet, when used out of France, it generally becomes an unfruitful exotic, and degenerates into cold conventionalism. Its blossoms invariably die in crossing the English Channel, and when imported to this side of the Atlantic there is little left of it but branches and withered leaves.

English architecture, on the other hand, is still groping after a fit type of national expression. If in France, under the patronage of government, there is a living style consistent with national traditions, a style still to a certain extent receiving accretions from the spirit of the times, thus serving as an index of national character, in England, without official guidance, liberty of thought is unrestrained except by the unrecognized influence of custom. The result is that the elements of design, which are repressed by the tyranny of a refined scholasticism on the other side of the Channel, find the fullest expression, while the study of detail and ornament, to which French genius has been compelled to confine itself, is essentially wanting. Thus, English architecture abounds in picturesque, romantic, and religious thought. Indeed, through these sentiments alone, it has occasionally succeeded in entering into the difficult regions of noble architecture; but, with certain exceptions quite rare enough to prove the rule, as in the Banqueting House at Whitehall, in St. George's Hall at Liverpool, in some of the earlier work of Barry in the club houses of London, and in some of Thompson's Greek work in Edinburgh and

Glasgow, English essays in classic types of architecture have hitherto been controlled rather by a Philistine conservatism and common sense, than by poetic feeling and inspiration. The orders have been used rather as an inflexible geometrical expression than as a language of art. In fact, the English have failed where the French have suceeeded; they have succeeded where the French have failed. But the English failure is the more disastrous to the rank of English architecture because its vain attempts to vitalize and nationalize the classic formula have been frequent enough to constitute a characteristic feature. On the other hand, the severe academic training of the French architects has preserved them from conspicuous error in any branch of picturesque or romantic effects which they have attempted. The French, in short, cannot be ungrammatical; but there is hardly a street in a modern English town which is not full of offenses against correctness. Notwithstanding this, the history of modern architecture in England, though its condition of artistic liberty has never given it its Augustan era, to correspond with that of Shakespeare or Addison in literature, attracts our interest and claims our sympathies to a greater degree than that of any other nation. It abounds in episodes of ingenuous and gallant effort, now in one shape and now in another, sometimes merely for the sake of change, and sometimes for the sake of principle. Not long ago, as has already been noted in a previous essay, it became necessary, in the interest of peace and quiet, to exclude by formal by-laws the discussion of the relative merits of classic and of Gothic art in the societies. The profession in England was divided into hostile camps by an irrepressible conflict of architectural principles. The Gothicists, by the chance of war, gained the day, and held the field undisputed for some forty years. This warfare, however grotesque it may now appear to us, who, by conviction and not by indifference, have become catholic, bears witness to a sincerity and zeal on pure questions of principle in art, which are unparalleled in history. Indeed, these are among the best fruits of liberty.

The main characteristic of modern English architecture consists in its series of revivals. In the absence of academic taste, guided by official schools, the architect is under the dominion of a prevailing fashion, which, while it lasts, is as powerful as if promulgated by an edict of government. He aims less to please with old forms than to astonish with new. Any strong mind or hand in the profession fortunate enough to make a happy revival of a style or phase of architecture, which had until then been laid aside and forgotten, establishes a starting-point for a host of young imitators, who at length constitute a school, numerously and enthusiastically followed; and thus a fashion in design takes possession of contemporary art and has its run through a course of years, until some other guiding spirit awakens a new revival and makes a new fashion, which succeeds until its capacity for producing novelty has been exhausted. The movements which have had the most enduring effect are the Greek revival, the modern Italian revival, the Gothic revival, the Queen Anne revival, the free classic revival; and now we are awaiting the manifestation of some master spirit who will turn the development of English thought, fatigued by these fruitless experiments, into some new and unexpected channel of archæology, or, perhaps, by a new theory of esthetics as applied to structure, direct it at last into some really modern and more intelligent field of effort in art.

It is a peculiarity of these successive experiments that they are revivals of completed systems, of forms incapable of further progression. Viollet-le-Duc, as I have once before noted, wisely remarked that a prosperous career in art can start only from primitive types, of which the powers of development are unexhausted by use. The practical effect of the revival of a style which has had its era of glory and its associations of history is to give to the architect an opportunity to exhibit his ingenuity in adapting old forms to new uses, and to display a facility of quotation which is often mistaken for genius, but which is really little more than memory, cultivated and effective indeed, leading him "to affect impressions which he does not feel," but not touching the springs of life in art. Meanwhile, the public are interested in it simply as they are interested in any other new fashion, not because it has in it the healthy breath of life, but because it is in vogue, and has been made reputable by architectural usage. Few architects have courage or force enough not to follow this usage. They are bound by it hand and foot while it lasts, and its powers are tested and strained to the uttermost limits by being forced into service often uncongenial to its natural capacity.

It is very noteworthy that the greatest of the English revivals, that of mediæval art, had its basis in an awakened conscience. Pugin and Ruskin preached the gospel of this revival; they asked for a return to the era of truth in art; they asked that architectural expression should be controlled by structure, and that decoration should follow the methods of nature. The Gothic revival is the only instance in history of a moral revolution in art. It is but fair to observe, however, that its long continuance in English practice may be partly attributed to that antiquarian spirit which has kept the architects busy in preserving and restoring the historic monuments of mediæval England. On the other hand, the revival of the style, called of Queen Anne, was a revolution effected largely under the influence of a literary sentiment. If the genius of Norman Shaw struck the first blow, the genius of Thackeray gave the movement inspiration and character. Both revivals were patriotic, and would have been impossible if not associated with phases of English history. But neither conscience, nor historic sentiment, nor patriotism, can make art; they can give character and variety, they can supply *motifs*, they can minister to emotions and inspire poetry, but they cannot make a style. Hence these English fashions, which have had loud and sometimes effective imitations in the practice of American architects, have apparently had no permanent influence in improving the practice of architecture either in England or here. They have made *dilettanti* among the public and *virtuosi* among the architects, but they have not created artists. There is plenty of archæology, but no inspiration, in an architectural fashion which is hampered by the necessity of strict conformity. "Creation and preservation," said Sergeant Troy, "don't do well together, and a million antiquarians can't invent a style." The Gothic revival gave opportunity for innumerable experiments; for many years it preoccupied the minds of the Anglo-Saxon people, and was ingeniously and sedulously adapted to every possible habit of modern life which could be ministered to by architectural forms. But this adaptation, though the writers have called its results "Victorian Gothic," did not advance Gothic art one step towards the creation of a modern English style, because it did not develop any capacity for expansion in this new service. Its potency for new expressions was speedily exhausted without satisfying the requirements of common sense or meeting the practical conditions to which it was applied. It was replaced — not, it is important to note, through any process of logical succession, — by the accident of the Jacobean, or Queen Anne, or free classic revival, which at every point was an offense to the architectural morality engendered by the preachments of Pugin and Ruskin, and possibly a result of them, as the license of Charles II. naturally followed the rigid Puritanism of the Commonwealth. This revival also is proving unsuccessful, because the capacities of the type had been already exhausted before the revivalist made his first quotation. Neither of these types permitted expansion or progress; and the rehabilitations of them have proved to be merely sterile incidents in the history of modern architecture. They have not, so far as we can see, furnished to the English civilization of the nineteenth century any fitting and adequate architectural expositions.

These revivals, as I have said, found a large and by no means an unintelligent expression in the United States. But the national genius of our architects and their freedom from the tyranny of historic precedent encouraged them to a far wider range of experiment in architectural forms. Out of these experiments hitherto there have as yet come no definite promises for art. But the higher education of the architect and the more exacting demand for studied and grammatical expression have in these latter days, especially in those parts of the country where the higher civilization obtains, already practically supplanted the illiterate products of our earlier conditions with better work. They have succeeded in establishing a higher technical standard of performance without loss of that quality of intelligent liberty in which lie our greatest hopes and our largest promise. The service of our architectural schools is already amply justified by its results. Their graduates have spread abroad in Western as well as Eastern States, and wherever they have had opportunity of practice they have sown good seed, and are steadily rendering obsolete the normal American types of raw and undisciplined invention, of audacious exaggeration and caprice. This is the first and most wholesome step towards rational reform. We have had good practice and experience in following the English fashions, but here their reign has never been undisputed. By the entire absence of local traditions; by the entire absence of monuments more ancient than those which we call " old colonial " (which we are recognizing for a little while in our practice to the extent of its limited but respectable capacity); by the entire absence of any official prejudice, of any venerable conventionalities, of any national system of instruction in architecture, we are left in a condition of freedom which is fatal to art while we are ignorant, but capable of great developments when we are educated. In regard to the use of precedent we are essentially eclectic and cosmopolitan. But education is enabling us to accord a proper degree of respect to the formulas and traditions of the Old World, to avail ourselves of them without bias, and to use them with a freedom which is becoming characteristic of our work. We are in position to profit by conventionalities without being bound by them. If our heritage of liberty has made us impatient of academical discipline, it has made us peculiarly hospitable to unprejudiced impressions of beauty and fitness. Our national offense has been license and insolent disrespect of venerable things, arising from want of appreciation and ignorance. We have carried experiment and invention in matters of design further than any other people. We are, as a new nation, a nation of builders. No part of the history of civilization is so various in its architectural expressions, good, bad, and indifferent, as that of the American people. In quantity, certainly, we have in a given time accomplished more in this field than any other people. Our distinctive practical necessities, our mechanical genius in the matter of building-appliances, the nature of our building-materials, the exigencies of climate, the characteristics of social life, and the invention of new systems of structure, like the braced and riveted steel frame, enveloped by terra cotta or by masonry of brick or stone as a fire-proof, have created certain corresponding distinctive qualities in our architecture; but they have not established as yet anything approaching that coherent body of architectural forms which constitutes a style.

The architect, in the course of his career, is called upon to erect buildings for every conceivable purpose, most of them adapted to requirements which have never before arisen in history. Practical considerations of structure, economy, and convenience preoccupy his mind, and his purely and conventionally architectural acquirements are subject to frequent eclipse in practice. His great architectural models often give him no hint, and stand too far apart from modern sympathies and use to serve him for inspiration and guidance. Railway buildings of all sorts; churches with parlors, kitchens, and society-rooms; hotels on a scale never before dreamt of; public libraries, the service of which is fundamentally different from any of their predecessors; office and mercantile structures, such as no preëxisting conditions of professional and commercial life have ever required; school-houses and college buildings, whose necessary equipment removes them far from the venerable examples of Oxford and Cambridge; skating-rinks, theatres, exhibition buildings of vast extent, casinos, jails, prisons, municipal buildings, music halls, apartment houses, and all the other structures which must be accommodated to the complicated conditions of modern society, — these force the architect to branches of study, to which his books, photographs, and sketches give him little direct aid. Out of these eminently practical considerations of planning must grow elevations of which the essential character, if they are honestly composed, can have no precedent in architectural history.

On the other hand, in the days when definite historical styles may be recognized, architecture was displayed in a comparatively limited class of structures; thus, in the thirteenth century there were ecclesiastical buildings, a few town halls, and feudal castles built for defence. This limitation kept practice within manageable boundaries and preserved the style from aberrations and perilous experiment.

Even though a prevailing fashion or revival may give a color of unity to contemporary buildings erected under the conditions which I have described, what wonder if there is the most perplexing variety of architectural expression? What wonder if the superficial critic, seeking for a characteristic type and finding none, cries out that art is dead, that there is no American style? What wonder if he decries the American architect as a creature without convictions? In fact, the canons of criticism, which guided the opinions of our forefathers under narrower conditions, and which led them to pronounce judgment with the formulas "correct," or "not correct," are no longer applicable. The art which of all the fine arts is the only one dependent on practical considerations must be a free art, from the nature of the case. It cannot be confined within the bounds of any historic styles and remain true to its functions; it cannot meet the requirements of modern life in a strait-jacket of antiquarian knowledge and archæological forms. The functions of the critic have become far more difficult, and require a far more catholic, unprejudiced, and judicial mind, a far wider range of knowledge and sympathy, than could possibly grow up under the teachings or examples of Vitruvius or Palladio, Philibert de Lorme or Sir William Chambers, Sir Christopher Wren or the brothers Adam, Pugin or Ruskin, or any other prophet or expounder of ancient principles, with their rigid doctrines of exclusions and their exact formulas of practice in design. It is an era of experiment and invention, of boldness and courage. Conscientious fidelity to style in the merely archæological sense no longer leads to great and successful achievements, because the requirements of modern buildings are far beyond its capacity. The narrow city façade, crowded with necessary windows, elbowed by uncongenial neighbors, restricted by municipal regulations, with story piled upon story, and the whole hanging over a void filled with enormous sheets of glass, the *bête noir* of architectural composition, is a defiance to all rule and precedent in art. Iron or steel construction can be adjusted neither according to the elegant precepts of Vitruvius nor the more elastic principles of mediæval art. At every step the architect is confronted by problems which cannot be solved by the suggestions of his library and sketch-book. He is compelled to employ new devices, to invent, to reconcile incongruous conditions, to strain the conventionalities of architectural design beyond their capacity, to produce new things. The result, in the absence of a wise and thorough architectural training in the fundamental principles of art, is confusion of types, illiterate combinations, an evident breathlessness of effort and striving for effect, with the inevitable loss of repose, dignity, and style. The practitioner becomes reckless of rules, and, despairing of being able to please, he aims to astonish.

Under these conditions, a new style of architecture — a style in the sense of the great historical styles, as those of the Greek, Roman, Byzantine, Romanesque, mediæval Saracenic, and early Renaissance periods — is impossible. But good architecture *is* possible. The progress of architectural knowledge has already begun to enable us to have our own revivals, and the experiments we are trying in this respect, being free from the prejudices of patriotic sentiment, which I believe to be a serious hindrance to the advance of English art, are curious and not without promise.

Among these revivals, that of the Romanesque forms of Auvergne, in which the vigorous round arches, the robust columns, the strong capitals, and the rich but semi-barbaric sculpture are tempered by reminiscences of the finer Roman art, is at the moment the most interesting and perhaps the most promising. The most powerful and imposing personality that has as yet appeared among the architects of America gave to this monument an initial force so great that now, after twenty years of experiment, it continues to hold its place as the most characteristic national manifestation of our architecture. The late Henry Hobson Richardson, who died, lamented, in 1886, in the midst of an exceptionally brilliant career, studied this especial phase of historic art with singular intelligence; but he poured into this antique mould such a stream of vital energy that the old type was transformed in his hands, and he indicated the direction of its modern development. In every city of the republic fully nine tenths of the new work in architecture shows in some degree the influence of this vigorous style. It has the advantage of being an early, unfinished and uncorrupted type, and it will be interesting to see in what direction and to what end its apparently unexhausted capacities will lead us, by the course of constant and intelligent experiment to which it is now subjected. But these experiments are often open to the charge of an affectation of barbarism and heaviness inconsistent with our civilization. They have, however, in the best hands, fairly broken loose from the bonds of archaic precedent in the style, and have already acquired new elements with a tendency to that delicacy and refinement which are necessary to satisfy modern culture, and to that elasticity which is essential to modern requirements. This revival will cease to be a masquerade, and will be in the healthy path of natural development, as soon as it begins to show a capacity for a more perfect adjustment to our material and moral conditions. Whether it will succeed in reaching this stage, or whether it will presently begin to fatigue by monotony and so fall into disuse, there is no question that the experiment is based upon sounder principles than that of any other now under study.

Certainly there is nothing in the slightest degree resembling it in the contemporaneous work of Europe. To take a larger view of the present attitude of American architecture, there is no doubt in my own mind that all the conditions are favorable to developments of the greatest interest. With thorough education, the future of our architectural practice is secure, and though "the American style" may never be realized, *style* will undoubtedly be a feature in our work, and will give it high rank in the history of art. It is a significant fact that we have ceased to copy the modern work of England or France, and that we are thinking for ourselves. The Columbian Exposition at Chicago has shown that, even in pure academic architecture, as taught in Paris itself, our work is not surpassed in modern times. But we are certainly not permitting this scholastic art to hamper in the least degree the development of an active inventive spirit in structure, such as no other country is exhibiting. The results of this spirit, as indicated in our high building and in our steel frames enveloped with fire-proofing, are already assuming distinctive architectural characters. Indeed, this spirit, more perhaps than any other influence, is defining our national architecture as an independent expression.

This essay began with a quotation explaining how the necessary equipment of the modern architect, his organized business appliances, his library, his prints, his photographs, his familiarity with the historic styles in all their phases of development, glory, and decline, his conscious æsthetics, — his study, in short, has predisposed him to insincerity and affectation, and prevented him from competing on equal terms with the creators of the great styles. There is another element of difficulty which it is important to note. Our predecessors, before the era of learning, had the good fortune to be sustained by public sympathy. Criticism they felt to be an active, incessant, and intelligent force. There could be no such thing among them as pedantry. The language of art was a common language which all understood; it was therefore used with force, correctness, and discretion, and broke naturally into poetry and song. The result was that the builder designed, not better than he knew, but under correction, and with the knowledge that his efforts would be appreciated at their full value; he was encouraged, surrounded, sustained, admonished, and taught by public opinion. If, in this atmosphere, he was a leader, it was solely by force of genius working in common paths. The people were all handiworkers. Fingers, brains, and heart wrought together at the forge, in the stone-cutter's yard, with the weaver's shuttle, at the jeweler's bench, with the carver's chisel, in the armorer's shop. The mind of every artisan was prompt to conceive, and his hand was quick to execute variations of form in his own craft, to give new interest and individuality to his work, however humble. It was natural

> Dass er im innern Herzen spüret
> Was er erschafft mit seiner Hand.

This constant exercise of intelligent invention made him sensitive to every expression of art, and he was not indifferent when his brethren, the masons or stone-carvers or painters or workers in wood or metal or glass, united their best individual handiwork in an architectural symphony. Under this pressure, with this powerful correlation of forces behind, the growth of definite architectural styles was inevitable.

It was no less inevitable, in the progress of civilization, that labor-saving machinery should take the place of handiwork; that artisans should become mechanics or tenders of machinery, in the products of which their best qualities of mind and heart could have no concern whatever. Thus the common artistic instinct, which, in the aggregate, constituted intelligent public sympathy in Athens, in Ravenna, in the Isle de France, does not exist in Manchester and New York. The architect is left to work out his problems alone, "with difficulty and labor hard," unsustained by appreciative recognition and criticism, buried in the study of ancient precedent and in the contemplation of theories of design. The fundamental motive of his work is changed, and the best results of his labors, more or less affected by the pale cast of thought, are set up in the public places in complete silence. Criticism, if any, has become indifferent and careless. The architect is neither praised for his good points, nor blamed for his bad. Few care whether they are good or bad. They are not understood. In this emergency the architects unite in societies, for the sake of professional fellowship and mutual encouragement. They form a caste apart; they are no longer the leaders and exponents of public sentiment and thought in questions of art; no longer agents in the development of a style. Not having the correction, the stimulus, and assistance of public knowledge and interest to keep them in a straight path of fruitful development, they make devious excursions in new fields, they "affect impressions which they do not feel," they masquerade in various capricious disguises, all with the hope of astonishing the ignorant and arousing the indifferent.

Under these circumstances, it is much to be feared that the advance of the profession in these modern days is, to a large extent, a progress of architecture into regions inaccessible to the public. The most intelligent laymen do not pretend to appreciate its motives or to comprehend its results. But without their aid architecture cannot advance.

The members of the profession have begun to say one to the other, Let us cease this process of fruitless sophistication; let us study how we can excite intelligent interest. This point once gained, we shall at length have established a standard of performance which, if advanced by us in our practice with wise persistency and honest, straightforward endeavor, may at length give us a public whose good opinion will be worth earning. We have discovered that mere caprice, mere novelty, will not answer. This may amuse the vulgar for a while, but it makes the judicious grieve.

When in our schools and in our practice we can succeed in cultivating a fine artistic feeling and in establishing really catholic ideals in design without falling into dilettanteism or into habits of mere imitation; when we can use our knowledge of good examples, modern and ancient, so that it will not betray us into quotations for the sake of quotations, into conformity for the sake of conformity; when we can work without caprice and design *reasonably*, so that every detail shall be capable of logical explanation and defense, without detriment to a pervading spirit of unity; when we can be refined without weakness, bold without brutality, learned without pedantry; when, above all, we can content ourselves with simplicity and purity, and refrain from affectations; we shall have conquered the indifference of the people, and shall have accomplished more than has yet been done in modern England with all its archæology, or in modern France with all its academical discipline, but we shall have done no more than should result from an intelligent use of our precious and unparalleled condition of liberty in art.

ARCHITECTURE AND POETRY.

These essays have several times incidentally referred to the close analogies existing between poetry and architecture, and have endeavored to show that the power which elevates the science of building into the domain of architecture and makes it a fine art is the same power which converts prose into poetry. This is a creative power, which refines expression with beauty of form, and illuminates reason with imagination. Harmonious proportion, metrical form, or, in other words, perfection of technique, though an essential element in both arts, does not in itself make architecture, though there have been in later times historical eras, as in the eighteenth century, when this noble art was so misunderstood that its metre, its technique, was the only quality which differentiated it from structure; and the public consequently ceased to interest itself in an art which made no deeper appeal to its sympathies. In like manner, the heroic form of versification, which prevailed through

DOORWAY OF ST. TROPHEME AT ARLES

the same era, was not unfrequently accepted as constituting in itself poetry. But architecture does not reach its highest estate until it is so infused with imagination and fancy — not undisciplined imagination or capricious fancy — that the ordered fabric delights the eye of every intelligent observer and excites emotions like a poem. It does not need technical knowledge to make its best appeals intelligible to any educated layman. Indeed, such knowledge, if not thorough, may be an impediment to just appreciation.

Under these conditions of close analogy, it might be supposed that the poets would, in the course of literature, somewhere recognize architecture and express its spirit in inspired words for the benefit of mankind.

The art is so closely connected with the development of humanity, so curiously in sympathy with the progress of civilization, so interwoven with the aspirations of the race, that, combined with its own intrinsic and infinite expressions of grace and fitness, one might suppose it would present peculiar attractions to the poets; that they would delight not only to describe and interpret its manifestations as they appear in historical monuments, but to imagine new forms fit to illustrate and adorn poetry's various moods. And yet, with one or two possible exceptions, whenever the muse does celebrate architecture, she seems to stoop from her high career, and to be afflicted with a paralysis either of the intellect or of the imagination, which leaves her unfit to express an intelligible idea on the subject.

It is well known that no two architects who have attempted to restore, on paper, the villa Laurentinum of Pliny, by following the detailed and elaborate description of it in his famous letter to Gallus, have succeeded in producing similar designs. Disraeli, in his "Curiosities of Literature," infers from this that it is idle to indulge in architectural descriptions, as they cannot succeed in presenting clear pictures, and that the pen should not intrude on the province of the pencil. But the question is not so much one of description as of interpretation. Architectural ideas and *motifs* excite in the minds of architects certain emotions, which are rarely shared in their fullness by the laity. But I hesitate to believe that it is impossible for the pen to convey to the public at least some part of these emotions. It seems unreasonable that certain defined capacities of delicate enjoyment should be in a condition of permanent and hopeless atrophy in the minds of the great mass of mankind. It is contrary to experience in other domains of human effort that there should exist in one art powers of expression which are incapable of some sort of intelligible exegesis. Of course, every fine art appeals to a certain range of faculties of appreciation which cannot be reached by other fine arts. Painting has something to say which sculpture cannot say; architecture has a message which cannot be repeated in music; and *vice versa*. It would seem that the inspired insight and passion of the poet should be able to sympathize with and to impart at least somewhat of the peculiar intellectual excitement created by all these arts. Indeed, poets have successfully attempted this in the case of painting and sculpture and music. But the art of the architect is hardly more technical than that of the musician, and surely the appeals of the former to the intelligence of mankind should arouse emotions as capable of expression by the art of the poet. If a monument of architecture is like a "song without words," it certainly touches the mind and heart as much as it moves the senses. The work of Callicrates, of Apollodorus, of Anthemius of Tralles, of the builder monks of Cluny, of the Abbé Suger, of Bramante, Palladio, and

Sansovino, of the other masters of architecture, ancient and modern, is no more a mere pedantic display of technique than the work of Mendelssohn, Beethoven or Wagner. The art is not merely conventional or academic; it is essentially an expression of humanity in its noblest and most God-like moods.

Under these circumstances, it is remarkable that this magnificent and inspiring art is generally reduced by the poets to the subordinate function of furnishing an indispensable background to the persons and movements of the poem, and is referred to with certain commonplaces of description which nearly always fail to suggest the really essential values of the theme, which betray either indifference or ignorance as to fact, and which often present impossibilities of form and structure. Sculpture, painting, statues, tapestries, all often receive worthy recognition in verse, but the noble shrine which incloses and protects them, for which they were made and of which they are a part, is passed by with a conventional epithet, conveying to the mind no recognizable image. In fact, the intrinsic qualities of architecture seem, for the most part, to be invisible to the poets and inaccessible to their sympathies. When they refer to a monument of this art, it is generally to recall some historical association or incident connected with it, to draw an inference, to point a moral or adorn a tale. They do not seem to realize that its pilasters or buttresses, its base and cornice, its windows and doors, its panels and stringcourses, its columns and arches, have assumed shape and character coincidently with the progress of mankind; that these features can be interpreted as demonstrations of humanity and as evidences of civilization, all highly idealized and converted into visible poetry; that their ornaments of sculpture are a re-creation of the works of the great Creator, reflections of nature, slowly developed in types and conventional forms by the action of the human mind through centuries.

The emotions aroused in the mind of an intelligent expert by the contemplation of a work of pure architecture, in whatever style and of whatever race, are necessarily complex and difficult to describe; but I am persuaded that the quality and keenness and scope of these emotions are such as have been awakened by no other of the fine arts. To the "capable eye" there is, in the first place, the charm of repose, which includes almost all the virtues of design. Then follow the gracious and caressing appeal of technical harmony and grace in outline and proportion, in symmetry or balance of parts, in color, texture, detail, and distribution of ornament; the pleasing evidences of scholarship without pedantry, if the work is modern, of the intelligent study and adaptation of historical styles to modern use, of reserved power, of the absence of affectation or caprice; the just subordination of the personality of the author to his theme; the skillful adjustment of means to ends; the perfect agreement between construction and decoration: and, in certain cases, the glad recognition of the audacity of genius in breaking through the trammels of convention, and creating a surprise which does not offend. In the second place, outside of technique, the student is moved, in the contemplation of an historical monument, by its poetic suggestions; by the effect of national or local spirit on the treatment of outline and detail, and of that unconscious but inevitable imprint made by contemporaneous political, religious, social, or commercial conditions, which differentiates an architectural achievement from any other work of fine art, and makes it an evidence in the history of civilization. He has learned that the architectural monument is saturated with humanity; that it contains the essential spirit of history; and that even a Grecian Ionic capital, for instance, the decoration of a Roman frieze, of a Gothic spandrel or capital, or of the panel of an Italian pilaster of the fifteenth century, is a highly figurative image of a phase of civilization. Every movement of their lines, all their combinations, their various methods of presenting natural forms, are eloquent to those who can understand them. But all these remoter suggestions are conveyed to the mind by implication, by figure or symbol, as in poetry, and not, as in prose, by direct statement. The language of architectural forms is one of infinite artifice; it is born of traditions, is shaped by conventions, and speaks in parables and apologues, which are patent only to those who have studied the growth of thought in the development of its signs and symbols. This language is not so much in construction, harmonious and metrically ordered, as in the decoration of construction. It becomes articulate in ornament, whether, like that of Egypt, its apparent *motif* is a remote paraphrase of the lily and papyrus; or, like that of Greece, a highly conventionalized and chastened apotheosis of the acanthus, the honeysuckle, and the seashell; or, like that of Rome, an ostentatious, opulent, sensuous development of the Greek forms; or, like that of the early Christians, a spiritualized reminiscence of the conventions of Greece and Rome, and a new creation of the flowers and fruits of nature; or like the mediæval ornament, an embroidery of cusps and crockets, and an artificial adjustment of natural forms to the rhythm of architectural order and harmony; or, like the ornament of the Moslems, an intricate but orderly tangle of geometrical lines; or, finally, like that of the Renaissance, an elegant profusion of wreaths, garlands, medallions, and emblems, with flowing stalks of artificial foliage, mingled with human figures and chimæras, all created and arranged to make symmetry more beautiful. These conventions were used merely to ornament construction, with no deeper object in view; nevertheless, the spirit of the times in which they are executed unconsciously gives them a peculiar character and significance.

Much of the sentiment, many of the emotions, to be derived from architecture are of course enjoyed by the layman according to the degree of his natural sensitiveness to impressions from works of art, and according to the liberality of his education; but it is apparent that to his complete comprehension of the full meaning of an architectural monument there is needed an interpreter, who can not only feel it, as an expert, in its evident and remoter meanings, but who, if possible, can analyze and demonstrate it, thus opening to the world this most fertile and most delicate source of intellectual and emotional delight.

We would first look for such an interpreter among the poets, who express their ideas with words and figures, symbols and allegories, just as the architects express theirs with a language of forms, compact of ideals and conventions, developed in the progress of mankind by a similar process of creative imagination; but apparently no poet, even in this century, the most inquisitive of all in the history of our race, has tried to comprehend the best thought which the sister art expresses, although, as we shall presently see, two or three of our own time seem to have shown that the obstacles are not insurmountable, when the spirit is willing and the mind informed.

The old metrical romances, like those of Lydgate, the monk of Bury, or Piers Plowman, like the "Romance of Sir Degrevant," or Bradshaw's quaint translation in verse of "The Lyfe of Saynt Werburge," or the "Faerie Queene," contain occasional references to architectural effects, more or less fanciful, but indicating an intelligent basis of observation, and a certain appreciation of some of the characteristics of mediæval ornament. These references, however, deal, for the most part, with details of furniture and fittings, and, though they have proved mines of wealth to the antiquarian, never, even where they are most definite, convey to the mind a distinct architectural image, or touch upon the essential and vital qualities of the art. These mediæval singers belonged to a time when architecture was a living art, close to the hearts of the people, expressing the spirit and aspirations of the times with a peculiar directness, a poetic force, an unaffected sympathy, which should have been, and probably were, entirely intelligible to them; and yet the great monuments of that time inspired them with no line of adequate recognition.

Even Shakespeare, with his world-wide range of sympathy and his immortal intuitions, apparently is unaware of the real relations of this art to mankind. His almost divine imagination seems, in this one respect, to have no loftier vision than that common to the time of Elizabeth. As to the functions of art in general he more than once, in passages illuminated with poetic fire, anticipated without an effort the labored conclusions of modern æsthetics. But though his Duncan and Banquo, as they approached the castle of Macbeth, discoursed on its pleasant seat and on the martlets who built upon it coigns of vantage, and though, in other scenes, he managed by an accidental word or phrase to suggest architectural scenery and surroundings, he never dwelt upon or interpreted a single feature of this noble art, or presented a definite image of any of its historical manifestations. Apparently no impression was made upon his mind even by contemporary buildings. Bacon, on the other hand, had a very intelligent interest in architecture, and wrote of it with far more sympathy than any of his contemporaries; but Shakespeare made no use of the frequent opportunities of his dramas to refer to it, save once, very indirectly, in the Second Part of " Henry IV.: "

"When we mean to build,
We first survey the plot, then draw the model;
And when we see the figure of the house,
Then must we rate the cost of the erection;
Which if we find outweighs ability,
What do we then but draw anew the model
In fewer offices; or, at least, desist
To build at all?"

Even this, however, is a recognition of practical processes of building, and not of architecture as an art. This absence of adequate allusion may serve as another proof, if another were needed, that the two great Elizabethan names do not stand for one personality.

Milton, in his description of the Satanic hall of council,

"Built like a temple, where pilasters round
Were set, and Doric pillars, overlaid
With golden architrave; nor did there want
Cornice or frieze with bossy sculpture graven,"

intended to convey the impression of a splendid fabric, magnificently ordered, of awful grandeur and extent. His theme clearly demanded at this point a superb manifestation of creative power in a vision of architecture, such as the habitual pomp and majesty of his poetic diction would have delighted to set forth, if his knowledge of the art had extended beyond the pallid convention of a classic temple. In fact, all the literary knowledge of his time in respect to this art is summed up in the well-known passage in Thomson's "Liberty," written fifty-seven years after Milton's death: —

"First unadorned,
And nobly plain, the manly Doric rose;

ARCHITECTURE AND POETRY. 247

> The Ionic then, with decent matron grace,
> Her airy pillar heaved; luxuriant last,
> The rich Corinthian spread her wanton wreath.
> The whole so measured true, so lessen'd off
> By fine proportion, that the marble pile,
> Form'd to repel the still and stormy waste
> Of rolling ages, light as fabrics look'd
> That from the magic wand aerial rise."

It was from a body of pedantic Vitruvian formulas, dry and inelastic as a demonstration of Euclid or a rule of grammar, that all the architectural notions of the period were derived. It was a part of the elegant learning of the day to commit to memory the four orders as they were then understood. Every demonstration outside of these orders was barbaric. And yet we search in vain through the rhymed heroics of Dryden, Pope, and their imitators of the eighteenth century for an appreciative or intelligible idea even of a correct classic composition. When they attempted it, the result was a shapeless, disordered, heterogeneous mass; set to most harmonious verse, indeed, but hopelessly inharmonious in the image. Vanbrugh, one of the numerous fashionable gentleman-poets of the time, himself the architect of Blenheim and Castle Howard, is not inspired, in his own verse, to correct the ignorant incongruities of his contemporaries. When Pope, imitating

Chaucer in the scheme of his poem, and Milton in his architectural imagery, essays to present a poetic idea of the Temple of Fame, we have, in elegant and facile rhymes, an horrific intermingling of crude hints of Doric, Barbaric, and Gothic styles, which can convey absolutely no sane impression of structure or form in outline or detail. If the poet of Twickenham villa, in his insatiable greed for knowledge, had considered it worth his while to master the simplest elements of architecture, how readily could he have enshrined, in the elegant artificiality of his lines, a subject so much in sympathy with his poetic methods as a classic composition, with its ordered peristyle, its walls rich with color and incrustations behind the open screen of marble shafts, its pilastered pavilions and sculptured pediments, its decorations of statues and painting, and, over all, its storied dome! It is hard to conceive how such an imagination could be indifferent to a fact so poetic, so orderly, so easy to comprehend, so adjustable to the purpose of his verse.

In fact, the general insensibility to effects of art in the eighteenth century is one of the most remarkable phenomena in the history of the intellect. Gothic art in especial suffered from this eclipse of feeling. Its most magnificent monuments, the great metropolitan cathedrals of the Middle Ages, themselves poems in stone, were not only neglected during this long period, but despised, insulted, and misunderstood. When they were referred to at all, they were stigmatized as demonstrations of barbarism. They touched no responsive chord in the human heart until the modern romantic school arose, and Boisserée in Germany, Viollet-le-Duc and Victor Hugo in France, Pugin and Ruskin in England, restored them to the admiration and affection of mankind. Until then, through all those long years, to the poets as well as to the common herd, they uttered absolutely no word, and gave no breath of inspiration. To the literature of this time architecture was merely a series of stiff, unpliant formulas of classic art, without principles, only half comprehended, — a fetich to pedants, an enigma to the people. Since the enlightenment furnished by the romanticists, since the exposition by certain late writers of the theory and principles of the art, the sentiment of architecture has begun to penetrate the tardy perceptions of the poetic instinct; yet only in two or three instances has it received anything like an intelligent recognition.

Towards the close of the period of sterility, one strong, clear voice broke the long silence with strains which accomplished more for the recognition of architecture in literature than all other agencies combined. Among all the poets, Sir Walter Scott seems to stand alone in his thorough appreciation of the value of real architectural background and accessories to the interest of romantic verse. He used his archæological knowledge and his fondness for mediæval architecture with the skill of a practiced romancer and the sympathy of a poet. His example was the potent factor in the creation of that particular romantic school in English literature which followed him. But none of his imitators approached this mighty minstrel in the truth with which the characteristic details of chapel, castle, or abbey were made essential parts of his picturesque stories.

Abbotsford itself, the realization in material substance of Scott's architectural ideals, is but indifferent architecture ; it is at best but pinchbeck mediævalism. These ideals of structure, however, found much happier expression in his verse, the plan of which did not necessitate exactness of portrayal, much less attempt to interpret the intrinsic properties of his mediæval models. The architect is relieved to find that this one poet, at least, did not make nonsense of his buildings. Whether the scene of his poem was laid in Melrose Abbey or Norham Castle, or whether it took him to the Saxon monastery of " St. Cuthbert's holy isle," to the stronghold of Crichtoun, or to the towers of Tantallon, the wizard's touch was true. His poetic visions never betrayed an historical monument. The flow of his imagination was corrected and held in check by the rare quality of honest loyalty to the facts of architecture as he understood them. Even in his details of description, though he touched them but lightly, the architect recognizes the salient points of the style of the building which he celebrates. His archæological knowledge was ever sufficient to his theme, and in great part inspired it, as was the case with Victor Hugo in " Notre Dame." If the schools created by these masters had, with poetic penetration and sympathy, continued the investigations of romantic art so brilliantly begun, literature would have been enriched by a new light out of the past, and architecture would, in some of its phases at least, have become an open book instead of an undecipherable myth or hieroglyph, of which the interest to the world resides in its outward grace, and not in its inward meaning.

If we turn to the entrancing stanzas of "Childe Harold," whose pilgrimages included the contemplation of the great masterpieces of art in Greece, Florence, Venice, and Rome, and who dwelt in immortal verse on the Venus de Medici, the Dying Gladiator, and the Laocoön, on the Coliseum, the Pantheon, and St. Peter's, we find that the architectural subjects, among all the works of art, are alone, from the point of view of the architect, inadequately treated. Byron's active and virile genius, prompt to appreciate a few palpable points of outline, and to enlarge upon the historical and romantic suggestions connected with the subject, contents itself with these. His quick insight and his descriptive powers are sufficiently evident, but, though the scheme of his poem certainly invites him to employ them on the most august themes that architecture presents, he fails to touch the really vital points. We cannot but be thankful for what he deigns to give us, and regret his failure to complete the work which in each case he begins with such splendid promise. He lingers long with fine historical emotions and tuneful meditations on the Acropolis of Athens, but the Parthenon furnishes no other inspiration than a spirited denunciation of Lord Elgin for stealing the

Panathenaic frieze! The Erechtheum he does not see at all, nor the Theseion; but he does espy the few remaining columns of the Roman temple of Jupiter Olympius. He visits the Bosphorus, but cannot find the matchless dome of St. Sophia, from beneath which the arts of Christianity and Islam parted on their divergent careers. What a subject for his muse! He is magnificent, however, when he enters St. Peter's, and shows clearly enough that his poetic powers can grow colossal with the greatness of his theme, and can, when he pleases, express an architectural emotion; but the pagan art by which this Christian pomp is expressed, and all that this art, as developed in the great basilica, stands for in the history of the human race, have absolutely no recognition. So far as his architectural description or references are concerned, his words would apply quite as well to the hypostyle hall at Karnak or to the Cologne cathedral. He enters the Pantheon without noticing the portico (which, however, another and later poet, Arthur Hugh Clough, did see in his time); he observes within that the sole source of light is from one aperture, and he sees the altars and the busts, but there is nothing to show that this sole aperture is open to the sky, and forms the eye of the dome, whose vast coffered concave, itself an epic poem, appeals to him in vain.

Perhaps the attitude of Byron towards this art is revealed in the stanza of the fourth canto wherein, with lofty disdain, he refers to the students of sculpture : —

> "I leave to learned fingers and wise hands,
> The artist and his ape, to teach and tell
> How well his connoisseurship understands
> The graceful bend and the voluptuous swell:
> Let these describe the undescribable."

But surely, to leave untouched all the deep human meanings involved in the purely architectural points of the great monuments of which he sings; much more, to remain insensible to such points as appeals of art, betrays at least an astonishing indolence of mind. In fact, like most of his tuneful brethren, he was a mere impressionist as regards architecture. Like them, he had not patience enough to study the subject, nor cared to penetrate the veil of conventionalism which shuts out from casual view its richest and most potent significations. They flattered and caressed the blurred and imperfect images made upon their minds by these objects of art, and delighted the world with their unstudied reflections.

Rogers is another poet who, like Byron, wandered through the old lands of art in search of inspirations; but if Byron did, at rare moments, break into irrepressible panegyric when some one of these great monuments of human intelligence and aspiration forced itself upon his reluctant apprehension, Rogers ransacked all Italy for poetic emotions, but apparently did not see a building from the beginning to the end of his metrical career.

Wordsworth, with less fire than Byron, but with a far sweeter and more patient poetic instinct, at times seemed almost to enter the enchanted castle, and to arouse to life its sleeping beauty. No architect can read his forty-third and forty-fourth ecclesiastical sonnets, on King's College Chapel at Cambridge, without grateful recognition.

> "Vex not the royal Saint with vain expense,
> With ill-matched aims the Architect who planned —
> Albeit laboring for a scanty band
> Of white-robed Scholars only — this immense
> And glorious Work of fine intelligence!
> Give all thou canst; high Heaven rejects the lore
> Of nicely calculated less or more ;
> So deemed the man who fashioned for the sense
> These lofty pillars, spread this branching roof,
> Self-poised, and scooped into ten thousand cells,
> Where light and shade repose, where music dwells
> Lingering, and wandering on as loath to die ;
> Like thoughts whose very sweetness yieldeth proof
> That they were born for immortality.

"What awful perspective! while from our sight
With gradual stealth the lateral windows hide
Their Portraitures, their stone-work glimmers, dyed
In the soft checkerings of sleepy light.
Martyr, or King, or sainted Eremite,
Whoe'er ye be, that thus, yourselves unseen,
Imbue your prison bars with solemn sheen,
Shine on, until ye fade with coming Night!
But, from the arms of silence, — list! oh list! —
The music bursteth into second life;
The notes luxuriate, every stone is kissed
By sound, or ghost of sound, in mazy strife;
Heart-thrilling strains, that cast, before the eye
Of the devout, a veil of ecstasy!

"They dreamt not of a perishable home
Who thus could build."

There is another master who penetrated deeper yet behind the veil, and showed that he not only appreciated a work of architecture, but understood somewhat of the structural form through which its sentiment found expression. No architect could ask for a clearer picture than that presented by Browning's half-pagan Bishop when he ordered his cinque-cento tomb in St. Praxed's Church. The voluptuous Renaissance of the episcopal cenotaph suggests a definite image in shape and color, not only of the material object, but of the idea behind it. It is but a sketch, yet it is touched with the hand of a master, whose inspiration has behind it not only feeling, but knowledge.

Tennyson, in his "Palace of Art," gives us the merest phantasm, like Thomson's "Castle of Indolence," or like the temple in Shelley's "Revolt of Islam." These are all cloud-capped visions in a "pleasing land of drowsyhed," without foundation or tangible substance.

Lowell, too, in his "Cathedral," is beautifully vague, and, though his poem is rich with precious thought, it wanders from its theme, and misses nearly all those points of true Gothic design and sentiment which present themselves with inspiring suggestion to the imagination of every architect who knows and loves his cathedral of Chartres. He himself frankly says: —

"I, who to Chartres came to feed my eye
 And give to Fancy one clear holiday,
 Scarce saw the minster for the thoughts it stirred
 Buzzing o'er past and future with vain quest."

But we could not ask for a more exquisite sketch, as the work of an impressionist, than this of the exterior: —

"Looking up suddenly, I found mine eyes
 Confronted with the minster's vast repose.
 Silent and gray as forest-leaguered cliff
 Left inland by the ocean's slow retreat,
 That hears afar the breeze-borne rote and longs,
 Remembering shocks of surf that clomb and fell,
 Spume-sliding down the baffled decuman,
 It rose before me, patiently remote

From the great tides of life it breasted once,
Hearing the noise of men as in a dream.
I stood before the triple northern port,
Where dedicated shapes of saints and kings,
Stern faces bleared with immemorial watch,
Looked down benignly grave and seemed to say,
Ye come and go incessant ; we remain
Safe in the hallowed quiets of the past ;
Be reverent, ye who flit and are forgot,
Of faith so nobly realized as this.

. . . .
. . . . . I give thanks
For a new relish, careless to inquire
My pleasure's pedigree, if so it please,
Nobly, I mean, nor renegade to art.
The Grecian gluts me with its perfectness,
Unanswerable as Euclid, self-contained,
The one thing finished in this hasty world,
Forever finished, though the barbarous pit,
Fanatical on hearsay, stamp and shout
As if a miracle could be encored.
But ah ! this other, this that never ends,
Still climbing, luring fancy still to climb,
As full of morals half-divined as life,
Graceful, grotesque, with ever new surprise
Of hazardous caprices sure to please,
Heavy as nightmare, airy-light as fern,
Imagination's very self in stone !
With one long sigh of infinite release
From pedantries past, present, or to come,
I looked, and owned myself a happy Goth."

If, with such affluence of imagination and diction, Lowell had not been, as he confessed, " careless of his pleasure's pedigree," like the rest of the impressionists in verse and in art ;

if he had had more of the pre-Raphaelite qualities, which inspect and analyze the source of pleasure before attempting to portray, he would have interpreted this lovely mediæval enigma like a prophet.

Emerson, with a delicate and almost unequaled depth of poetic insight, touched, as it never had been touched before or since, one truthful chord in "The Problem:"

> "The hand that rounded Peter's dome
> And groined the aisles of Christian Rome
> Wrought in a sad sincerity;
> Himself from God he could not free;
> He builded better than he knew; —
> The conscious stone to beauty grew."

But he left his brief architectural strain too soon, and left how much unsung!

Longfellow's muse, it must be admitted, is indebted for some of its happiest imagery to his fine consciousness of the romantic sentiment in Gothic art. No European poet, born and bred in the shadow of cathedral or cloister, ever felt more deeply than this sweet minstrel from Maine, or expressed more tenderly, the emotions which a true poet should feel under these influences. But he never wrote an architectural poem, and it is very evident that he never studied and did not really comprehend the true Gothic, which he loved so much and which inspired so much of his verse, nor imagine the infinite lights and shadows of human life hidden behind the mediæval mask. If we read his lovely lines on his translation of the "Divina Commedia" of Dante, we may see that, when he touched upon architecture, he merely used it as a rhetorical image, secondary to a thought outside of the art. As an example of the delicacy and truth of his poetical workmanship under such limitations, I may be permitted to quote his second stanza : —

> " How strange the sculptures that adorn these towers !
> This crowd of statues, in whose folded sleeves
> Birds build their nests ; while canopied with leaves
> Parvis and portal bloom like trellised bowers,
> And the vast minster seems a cross of flowers !
> But fiends and dragons on the gargoyled eaves
> Watch the dead Christ between the living thieves,
> And, underneath, the traitor Judas lowers !
> Ah ! from what agonies of heart and brain,
> What exultations trampling on despair,
> What tenderness, what tears, what hate of wrong,
> What passionate outcry of a soul in pain,
> Uprose this poem of the earth and air,
> This mediæval miracle of song ! "

The architect can never forget that noblest of all the poetic tributes to his art which Longfellow puts in the mouth of his Michael Angelo : —

> "Ah, to build, to build !
> That is the noblest art of all the arts.
> Painting and sculpture are but images,

> Are merely shadows cast by outward things
> On stone or canvas, having in themselves
> No separate existence. Architecture,
> Existing in itself, and not in seeming
> A something it is not, surpasses them
> As substance, shadow."

The architect who completed for the Duchess Marguerite the church of Brou, and the sculptor who carved the tomb within, though they aimed to appeal to such refined intelligence as that of Matthew Arnold, and to touch with their art such sensitive hearts as his, studied their careful details and created their harmonies of form in vain so far as he was concerned. His lovely lines on this church and tomb have no recognition of the fullness of the message which these objects were intended to convey, and present no clear picture even of their apparent form. His delicate instinct, when confronted by the visible poetry in these monuments of art, felt no sympathetic thrill, and saw only the effect of the tinted light from the windows as it played upon the pavement; the shafts and groined vaulting of the church appeared to him only a "foliaged marble forest;" and his poetic eye discovered only "chisell'd broideries rare" and "carved stone fretwork" on the tomb, where were sculptured the two forms, —

"One, the Duke in helm and armour ;
One, the Duchess in her veil."

The competent interpretation which the "frozen music" of one art had a right to demand from the inspired insight of the other is vainly sought for in this beautiful verse. Here, as elsewhere, the sister arts remained strangers one to the other, and the real architecture of the church of Brou was invisible to one who should have been its oracle.

Byron's haughty disdain for the study of a work of art may be something more than a personal idiosyncrasy; it may represent the characteristic attitude of all his poetic brothers and sisters. But a mind familiar with this noblest of the fine arts, and trained to its practice, finds it difficult to condone this indolence or indifference of the tuneful choir. Of course it may be said that a poet need not be a geologist or a botanist to enable him to treat a landscape in adequate poetic phrase, and that, therefore, to celebrate justly an architectural theme, the equipment of an architect or of an archæologist is not necessary to him; that the "Tintern Abbey" of Wordsworth, for instance, is a beautiful and satisfactory poem as it stands, and that it would have been no more acceptable if, instead of the exquisite reflections which were actually incited in his mind by the neighborhood of that monument, it had inspired him with thoughts more germane to its intrinsic architectural and human conditions. In fact, it is because of these conditions, because it is a creation of the culminated and combined wisdom of mankind at the moment of its erection, and a poetic expression of the civilization of its times, that a monument of architecture has a different sort of interest from the works of nature, a significance which cannot be reached by a casual impression of some of its external effects. The "Tintern Abbey" of Wordsworth was not intended to be a poem of architecture. For the purposes of the poet, this building did not differ in value from a demonstration of nature; to him it was a mere mark of locality, inducing a certain range of thought because of association. An architectural poem on such a subject would be at least equally well worth writing: it would celebrate in poetic form the structural and decorative harmonies of the subject, and would enter into the feelings of the men who created it; it would reveal the deep significance of its individuality of character in form and detail; it would touch upon the human aspirations and passions unconsciously built into its walls, and would draw its inferences and lessons from these inherent conditions; it would be a poem of humanity, based upon one of humanity's most exquisite manifestations. Such a poem, apparently, has not been written.

Now, pondering these things, it has occurred to me to question whether the explanation of this silence of the poets lies in the fact that no expert has as yet shown the way to this region of difficult access, so that the inspired ones might at last find an entrance by following his footsteps, and gather there the flowers that so long have blushed unseen and wasted their sweetness in vain; or whether, after all, it may be impossible to describe architecture adequately in sympathetic poetic diction, avoiding technicalities, which would be the merest stumbling-blocks to inspiration, and to express, by the same medium, somewhat of its true sentiment and meaning. The answering of the questions seemed to be worth a somewhat hazardous experiment. To this end, quite conscious of the absence of the divine afflatus in my own composition, though with a lively appreciation of its results in others, and encouraged by the reflection that genius has been called the art of taking pains, and that patience is one of its most potent ingredients, I timidly, and with no exalted expectations, have ventured to try my 'prentice hand on

"Things unattempted yet in prose or rhyme."

For the sake of its simplicity, mainly, and because its capacities for the purpose seemed to be reasonably obvious and manageable, I have chosen for a subject a doorway in southern Romanesque, having in my mind, not an individual example, but rather a type; so that the characteristics of the style might not be subjected to the accidents, or limited to the idiosyncrasies, of a single monument. Perhaps, however, features of the porch of St. Trophême at Arles may have had a somewhat prevailing influence over the ideal which I have attempted to portray. Though of course the development of my thought has been materially embarrassed by the unfamiliar obstructions of rhyme, rhythm, and poetic diction, and my progress has been consequently slow, laborious, and plodding, quite without anything approaching what I understand to be meant by "fine frenzy," the performance of this self-imposed task has not been without somewhat of the "pleasure of poetic pains." The theme was at the beginning mapped out in cold blood, but I fancy that the form of the composition has forced it not only to overflow the original prudent boundaries of the argument, but at several points to take an unexpected turn of emotion or imagination, which I dare to hope may possibly be explained or condoned as the process of the evolution of prose into what may be called poetry. I am not at all sure on this point; but the process, I believe, if it has to some extent idealized the thought, or perhaps led it astray, has not betrayed the architecture, for the integrity of which I must be held responsible.

Possibly the method of presentation which I have employed may, in skilled and practiced hands, render architecture intelligible to those who, as Burke said, are ready to yield to sympathy what they refuse to description. At all events, if the results which mere plodding industry, under the impulse of long-cherished enthusiasm and corrected by a reasonable knowledge of the subject, has reached, may not afford a sort of pleasure to others, it may at least interest them, as, under the circumstances of its production, a curiosity of literature.

If it is remembered that in this modest

experiment I do not rashly pretend to compete with the poets, nor even to prove that the field, which I still believe to be rich in poetic thought and abounding in food for the imagination, is accessible to them, the obvious comment will not be made, that "fools rush in where angels fear to tread."

I am tempted, therefore, to release this child of painful endeavor from its secret place, with the pathetic inquiry upon its lips, "Have I a right to exist?"

THE CHURCH DOOR.

A STUDY IN ROMANESQUE.

TWICE four hundred years have borne
To this doorway, gray and worn,
Weary weights of grief and sin ;
Contrite, have they entered in,
And, beneath the arch of stone,
Laid their burdens down, and known
That to faith, whate'er betide,
The doors of heaven are opened wide.
For, with invitation sweet,
The pastoral Church, her flock to greet,
To fold, to comfort, and to feed,
This Portal Beautiful decreed.

The narrowing arch is deep and wide ;
Niched in its jambs on either side,
Shaft beyond shaft in ordered state
Stand on their solid stylobate,
Their leafy capitals upholding
Archivolt and fretted moulding ;

Arch within arch, with lessening leap,
From shaft to shaft concentric sweep,
Echoing inward o'er and o'er,
Inward to the vaulted door,
Every arch by subtle hand
Wrought with roll or bead or band,
Wrought with fillet or with fret,
Dentil, billet, or rosette,
While, between the sculptured rings,
Angel choirs spread their wings,
And, soaring as the arches soar,
With viol and with voice adore.

For the happy masons said,
As the radial stones they laid,
Truly wedged, with every joint
Loyal to the central point,
And by touch of chisel taught
Utterance of human thought, —
"Let the choral arches sing
Joyfully a welcoming,
Every one in concord fair
Moulded and attuned to share
By the cunning of the mason
In a solemn diapason,
While the great arch over all,
Silent, bears the mighty wall;
Silent, while its arch-stones deep
Under the sheltering label sleep,
And the corbel-heads intone
Vespers with their lips of stone."

Then with reverent hands they laid,
Deep in the archèd frame embayed,
Circled with immortal song,
Upon a lintel deep and strong,
A sculptured slab, to symbolize
Grace Divine to human eyes.

Oaken doors they hung below,
On forgèd hinges turning slow,
The rigid iron branching wide
With foliate growth from side to side.

Blessèd they who enter here !
For, upon the midway pier,
The gentle Mother, undefiled,
Bears on her breast the Holy Child,
And, born of superstitions old,
Consecrated types unfold
To purer meanings, and impart
Dignity to childlike art.
Ranged along the lintel stone,
Each like each, and all like one,
Side by side in sad debate,
The twelve apostles sit in state.
High on the stone where Grace Divine
Shows to mankind the sacred sign,
The blessèd Lord, in glory crowned,
Sits majestic, while around
His central throne, on either hand,
The four mysterious Creatures stand,
Ready to bear, with wings unfurled,
His great Evangel to the world ;
His hand, upraised in benediction,
Comforts pain and soothes affliction,
Ever blessing year by year
All who humbly enter here,
Saying at the open door,
Pax vobiscum evermore.

With craft by gray traditions bound,
The builder raised these arches round,
Developing in progress true
The ruder forms his fathers knew.
He built them strong with honest care,
With heart of pride he built them fair,

Prodigal of labor spent
In joyfulness of ornament;
Not yet by learning led astray
From nature's strong and simple way,
Not trained as yet to analyze
The gifts of God with questioning eyes,
Nor by sophistication cold
Made timid where he should be bold,
No fine restraint the builder knew;
Barbaric force to beauty grew
In types of unaffected form,
From the heart of nature warm,—
Prolific roots with strength innate
In future growths to germinate,
The perfect flower, yet unblown,
Hidden in the sheath of stone.

To him no rich, historic Past,
With strange ideals and visions vast,
Held bitter fruit of knowledge out
To tempt his innocence with doubt.
In narrow bounds his course was laid;
Not distracted, nor afraid,
Here he worked with earnest heart,
Nor knew his handiwork was art;
In images, as nature taught,
And not in learnèd words, he thought;
Carved as his fathers carved of yore,
But with a touch unknown before,
And kept his living art apace
With the progress of his race.

The sterile stones to life awake;
O'er the naked fabric break
Growths from ancient classic seed,
Acanthus, ovolo, and bead;
But the flowers of the field
Secrets to the carver yield,

And, new-created, play their part
In the symmetries of art.
Stem and tendril, bud and bloom,
Here an order new assume,
Trained to fit the builder's place
With artifice of formal grace.
All living things, by art transformed,
In this new creation warmed,
To new uses strangely grown,
Animate the bossy stone.
From these vital forces spring
Forms of prophet, priest, or king,
Scarcely wrought on nature's plan,
More of stone, and less of man,
Mystic types of faith and hope
Greater than the carver's scope,
Who with baffled art would trace
In outward form an inward grace;
Tall, attenuate, and still,
Like the niches which they fill,
Right and left, the door they keep,
Watching, while the ages sleep.
Here the carver's wayward tool
Breaks through order's rigid rule,
And grotesquely, as he works,
Humor gross with worship lurks.
Creatures of invention strange
Through the sculptured leafage range,
As with strains of music stirred,
By all ears but theirs unheard,
Moving rhythmic with the bent
Of structural line and ornament,
Pursue their sports or chase their prey
In a carver's holiday.

Tales of Scripture, legends old,
In the crowded caps are told,
Which, with leaf or figure, still
Under the abacus fulfill

In various forms their double duty
To bear with strength and crown with beauty.

Now, nature, with her soft caress,
Has stooped the carver's work to bless
With the mystery and surprise
Of her silent sympathies.
Centuries, whose mellow tones
Sleep upon these votive stones,
Have smoothed the threshold with the beat
Of their penitential feet ;
And age to art a grace has lent
Quite beyond the art's intent,
To conscious stone, to human thought,
A new interpretation brought.

The spirit, waked by patient art
In the quarry's sleeping heart,
Taught to repeat its lessons clear
On fretted arch and storied pier
In utterance beautiful, that finds
Quick access to lowly minds,
Whispers from the graphic stone
Solemn secrets of its own, —
Runes which, heard aright, betray
Stories of the ancient day ;
Tell what never scribe nor sage
Wrote on the historic page
Of arts and manners, and the place
Reached in the progress of the race,
When builders toiled, untaught but true,
And "builded better than they knew ; "
Tell, in some new exotic grace
In turn of leaf or chisel's trace,
In flower's shape or carver's thought,
The sources whence the people caught
The strength to rise from old to new,
From dark to light, from false to true ;

Between what nations far aloof
Commerce wove its peaceful woof ;
What Byzantine fires came
To set their smouldering arts aflame ;
How from Roman shrines was brought
Pagan wealth to Christian thought ;
What forces and what fates combined
To change the courses of the mind,
To mould the destinies of thrones,
While the carver carved the stones.

Still, as of yore, the arches throw
Over the door their sacred bow,
And with their invitation old
Gather the flock within the fold ;
Still to the flock in peace and rest
The builder's graphic art is blest ;
The vocal stones in concord sweet
Their ancient lessons still repeat ;
But what is hidden few may read
Behind the church's sculptured screed.
If the powers which shaped the fate
Alike of lowly and of great
Hieroglyphic record made
When these humble stones were laid ;
If acanthus and volute,
If this growth of leaf and fruit,
This new creative force, which brings
New life to all created things,
Kindling with its vital flame
The lifeless geometric frame,
Gathered in from race to race
Increments of time and place,
And the nations set their signs
In the carver's forms and lines ;
If the spirit that awoke,
'Neath the unconscious chisel-stroke,
Was the soul of history, —
Deep into its mystery

Let my new-world vision see,
When these doors unfold for me,
When upon this threshold-seat
Linger my expectant feet,
And the blessing on my head
From the lifted Hand is shed.

When, beneath far Western skies,
Seeds of this ancient art surprise
The children of a younger race
With blossoms of exotic grace,
While the vigorous germs retain
The virtues of their primal strain ;
When, in sweet and virgin earth,
They find a new and prosperous birth,
And, in spacious, strenuous air,
A growth more free, a bloom more fair, —
So may a strong and simple art,
Born in innocence of heart,
Unfold beyond the builder's hope
In purer line and larger scope,
And modern life and light fulfill,
With studied aim and conscious skill,
The promises which, all unknown,
Slept in the old prolific stone ;
So may a living art be freed
From the pedant's narrowing creed,
And the awakened age at last
Break its bondage to the past.

www.ingramcontent.com/pod-product-compliance
Lightning Source LLC
Chambersburg PA
CBHW032053220426
43664CB00008B/976